T0320606

Digital Transformation in a Post-COVID World

Digital Transformation in a Post-COVID World

Sustainable Innovation, Disruption, and Change

Edited by
Adrian T. H. Kuah
Roberto Dillon

CRC Press
Taylor & Francis Group
Boca Raton London New York

CRC Press is an imprint of the
Taylor & Francis Group, an **informa** business

First Edition published 2022
by CRC Press
6000 Broken Sound Parkway NW, Suite 300, Boca Raton, FL 33487-2742

and by CRC Press
2 Park Square, Milton Park, Abingdon, Oxon, OX14 4RN

ISBN: 978-1-032-07738-3 (hbk)
ISBN: 978-0-367-70958-7 (pbk)
ISBN: 978-1-003-14871-5 (ebk)

DOI: 10.1201/9781003148715

Typeset in Minion
by KnowledgeWorks Global Ltd.

Contents

About the Authors

Adrian T. H. Kuah A research leader at both The Cairns Institute in Australia and Centre for International Trade and Business in Asia at James Cook University, Adrian Kuah has published close to 70 papers in leading scientific and business journals, such as *Science of the Total Environment* and *Journal of Business Research*. He has represented both the British and Singaporean governments at various fora, such as conducting policy dialogues, leading in standardisation projects and advising on competitiveness and innovation policies in New Zealand, Singapore and Ireland. To date, he has received close to $900,000 of external funding towards his research. In 2013, he received the accolade of Professor of the Week in the *Financial Times* newspaper.

Roberto Dillon Born in Genoa, Italy, and based in Singapore since 2005, Associate Professor Roberto Dillon is the author of five well-received books published by AK Peters, CRC Press and Springer. A speaker at many high-profile international conferences and events, including GDC in San Francisco and TEDx in Milan, he is currently the Academic Head for the School of Science and Technology at James Cook University Singapore, where his research focuses on game design, history of technology and cyber security. In 2013, he founded and has since directed the very first Museum of Video and Computer Games in South East Asia.

Contributors

Abhishek Bhati
Professor and Campus Dean
James Cook University Singapore
Singapore

Agostino G. Bruzzone
Professor and STRATEGOS
    Council Chair
University of Genoa
Genova, Italy

David J. Calkins
Regional Managing Partner
Gensler
Singapore

Guglielmo De Gregori
Advisory Board Member
VIGAMUS Foundation
Rome, Italy

Mark Esposito
Professor
Hult International Business School
Cambridge, Massachusetts

Arizona State University
Phoenix, Arizona

Esther Fink
Learning Technologies Specialist
James Cook University Singapore
Singapore

Simran Grewal
Associate
KPMG Singapore
Singapore

Chang H. Kim
Postgraduate Research Fellow
The Cairns Institute
Cairns, Australia

Alessandro Lanteri
Professor
Hult International Business
    School
Dubai, UAE

and

ESCP Business School
Turin, Italy

Stewart Lockie
Distinguished Professor
James Cook University Singapore
Singapore

Director
The Cairns Institute
Cairns, Australia

Paul Lothian
Director, Cyber
KPMG Singapore
Singapore

Marina Massei
Simulation Team
University of Genoa
Genova, Italy

Victor Mills
Chief Executive
Singapore International Chamber
  of Commerce
Singapore

Malobi Mukherjee
Lecturer
James Cook University Singapore
Singapore

Daryl Pereira
Partner and Head of Cyber
KPMG Singapore
Singapore

Marco Accordi Rickards
Executive Director
VIGAMUS Foundation

and

Professor
University of Rome Tor Vergata
Rome, Italy

Micaela Romanini
Vice Director
VIGAMUS Foundation
Rome, Italy

Christopher Rouen
Research Officer
College of Healthcare Sciences
James Cook University Singapore
Singapore

Michael S. Tomczyk
Former Managing Director
  (retired)
Mack Institute for Innovation
  Management
The Wharton School
Philadelphia, Pennsylvania

Terence Tse
Professor
Hult International Business School
London, United Kingdom

Vihangi Vagal
Cyber Security Consultant
Deloitte
London, United Kingdom

Christopher Warren
Managing Director
Strategy & Consulting
Accenture
Singapore

Caryn West
Deputy Academic Head
College of Healthcare Sciences
Dean of Research
James Cook University Singapore
Singapore

Melissa Wingard
Senior Commercial Technology
  Lawyer
Phillips Ormonde Fitzpatrick
Melbourne, Australia

Caroline Wong
Senior Lecturer and Associate
  Dean (Teaching & Learning)
James Cook University Singapore
Singapore

Adrienne Yuen
Sector Specialist
Standards Council of Canada
Ottawa, Canada

Introduction

Standing on the Shoulders of Giants: Sustainable Innovation, Disruption and Change

'Standing on the shoulders of giants' is a metaphor used frequently in signifying that intellectual progress is made by means of extending the knowledge and understanding gained by major thinkers who are giants in their own rights.

'If I have seen further,' Isaac Newton wrote in a 1675 letter to fellow scientist Robert Hooke, "it is by standing on the shoulders of giants."[1] It is easier to learn from geniuses like Newton and Hooke, but even their work and ideas had been built upon the work and ideas of others before them.

Ideas for innovation and change have to come from someone or somewhere. No matter how unprecedented a work seems, dig a little deeper and you will find that the creator stood on someone else's shoulders. With each preceding observation, idea or reflection, one could see a little further. In turn, our thoughts and ideas would form the basis of the knowledge and progress for future generations. We would argue that 'Standing on the shoulders of giants' is a necessary part of innovation, disruption, and change.

The year 2020 started with all the usual promises of further technological advancements and a global economic rally that showed no signs of slowing down. Little did we know that, in a matter of weeks, our usual lifestyle and established habits were to be completely disrupted and we had to rethink the way we live and work together. The sudden rise of the COVID-19 pandemic forced all of us, across all industrial and academic fields, to go back to the writing board. We not only had to reimagine how to keep innovating, but also how to do old things in a new way.

This book captures how giants of our time – industry giants and thought leaders in their respective fields – across several geographical regions and domains faced this time of crisis, how they analysed the new, fast-evolving scenarios around them and finally found the right inspiration to move forward by recommending best practices for all of us to follow. The book brings together a mosaic of great ideas, observations and reflections from these giants on the importance of digital transformation in a post-COVID world by reliving some of their experiences.

Our lives were disrupted in many ways as the pandemic had a truly dramatic domino effect with far reaching impacts. As business models were disrupted and businesses closed around the world, we will see how leaders had to step-up and use technology as a mean to rethink their operations from the ground-up. Indeed, digital technologies permeate all aspects of modern life: mobility, health, education, work, communication, and so on. The ability to integrate these into our daily routines has literally become a matter of survival and we can do this only by remaining flexible. The ability to pivot towards new solutions as new trends emerge and force us to shape new strategies is indeed a common theme across the book.

At the same time, we also want to stress the importance of remaining vigilant and up to date with the latest technological trends as such drastic transformations do have hidden dangers, such as cyber security threats and privacy concerns.

The post-pandemic world will not be completely controlled by technology, though, as we still need to appreciate and, sometimes, rethink our lives in a fully holistic way. We should never forget about the spaces we live in and be sure that our living environment is still conducive to a healthy lifestyle, something we can easily fail to think of if we start living in a complete symbiosis with our computer screens.

A new idea on sustainable innovation, disruption and change is only possible because a giant opens a new door by introducing their idea, observation or reflection. They in turn opened new doors and expanded the realm of possibility. Science, art and other technological advances are only possible if someone else has laid the groundwork; nothing comes from nothing. Scientists like Newton and Hooke advanced science because of preliminary discoveries that others had made. With diverse contributions stemming from public health, technology strategies, urban planning and sociology to sustainable management, this volume is articulated into four distinct but complementary sections of People, Process, Planet and Prosperity influencing the post-COVID world.

PEOPLE

The ***People*** section discusses the basis of human wellbeing, mental health and the roles of leaders in business and society during and after the COVID-19 pandemic, as well as changes to industries, such as digital currencies, telework, retail and online shopping. Our thought leaders, such as Michael S. Tomczyk and Victor Mills, reflect on their observations on trends, leadership and challenges arising from the crisis, particularly in the United States and Singapore. In this context, telehealth proves to be one winner for mental health services in a post-pandemic world. Caryn West, Professor of Nursing, and Christopher Rouen at James Cook University examine the importance of telehealth mobilised by the COVID-19 pandemic, together with the impact and implications of large-scale infectious disease outbreaks, arguing why COVID-19 has the potential to negatively impact mental health and wellbeing. Research suggests that COVID-19 has resulted in an increase in the prevalence of psychological distress within general populations and an intensification of psychological distress among individuals. Additionally, the numerous uncertainties caused by COVID-19 are predicted to increase suicide rates and mental disorders associated with suicide in the foreseeable future. The importance of telehealth then cannot be understated. Although it will be important to invest in mental health now to assist individuals in these challenging times, it is likely that the effects of COVID-19 will linger slightly longer, highlighting the need to minimise the echo effects of COVID-19 into the future.

Victor Mills, Chief Executive of the Singapore International Chamber of Commerce, reflects on ethical behaviour in the business world and selfless leadership exhibited by some leaders. The COVID-19 pandemic has increased peoples' appreciation of authenticity, the ability to communicate with empathy and to act decisively in the best interests of the whole community. One positive aspect arising from the pandemic, Mills contends, is that business leaders are more aware of their own mental health and that of their team.

Companies and countries failed during this pandemic because of a failure of leadership in business and in public life. Mills examines the unintended consequences of globalisation and international trade in Singapore, and the importance of collaborative leadership. He contends that "selfless, constructive, collaborative leadership" is imperative. In this crisis, businesses need to work together to win the trust of their teams, customers, investors, and communities to transit into a newer economy from "wasteful linear to circular models of production and consumption".

Humans should not shirk on their responsibilities and accountability as leaders – acting for the common good. Too many of us abdicate leadership to others without recognising our responsibilities as citizens to participate in the life of the communities. Many reject the notion that they have any power to influence elections, reflects Mills. He further argues that during this pandemic, a large part of individual leadership is caring about positive outcomes for the community and for the country, as well as for the individual.

Michael S. Tomczyk, the technology pioneer responsible for bringing computers in the home of millions while at Commodore in the early 1980s and later a managing director of the Mack Institute for Innovation Management at The Wharton School, reflects on his observations in the United States, as businesses have to invent new ways of communicating, new ways of selling things, new ways to shop and new ways of running their operations.

One of the major behavioural changes accelerated by this pandemic was the shift from 'on schedule' to 'on demand' services, observes Tomczyk. Instead of accessing a product or service at a time or place determined by the seller, a buyer chooses when and whence to shop, with the item delivered directly to them, often with free shipping. This shifted temporal and spatial boundaries, which also affected the movie and entertainment industries. He further argues for the need of a cashless society and financial technologies. Advanced technologies are redefining the concept of money from paper bills and metal coins to digital currencies, he notes.

As a futurist, Tomczyk sees the continued rise of telework, 3D shopping, and online shopping. He notes that the pandemic has created a wide variety of chain reactions or 'dominoes' that have threatened and transformed our society. Supporting Mill's observation, Tomczyk points out the Yin–Yang tug of war between too much and not enough government control. His strongest takeaway from COVID-19 is that this crisis has given the world a wakeup call to better adapt to future crises.

The continued rise of digital servitisation and online shopping also threatens the retail industry worldwide – post-COVID, Debenhams in the United Kingdom and Robinsons in Singapore are examples of iconic stores that have joined a long list of well-known retailers called into administration. What lessons can future leaders learn from this turn of events?

Malobi Mukherjee, a futurist and research fellow from the University of Oxford and now lecturing in James Cook University, provides her perspectives on the traditional retail sector. Mukherjee proposes three retail scenarios for a post-COVID world that provide future safe spaces for retailers to develop resiliency. Her viable scenarios pertain to a hyper-tech world, an insular world, and a greener world, which entails real estate owners assuming the role of collaborative creators – a mutually beneficial nexus of new relationships and cooperative ecosystem.

PROCESS

The *Process* section investigates how workplace processes have experienced changes enabled by technology. Digital transformation took place in different sectors in a variety of ways, as more activities moved online and the definition and constraints of space and time warped. Nonetheless, digital transformation has its own unique risks and challenges, as we witnessed increases in cybercrime that affected governments all around the world and companies regardless of their size.

Christopher Warren, Managing Director, Strategy & Consulting at Accenture, helps navigate these new uncharted waters by discussing a technology roadmap to illustrate how different companies can pivot their businesses in the post-pandemic economy effectively. Indeed, technology has a major part to play in dealing with the pandemic and will continue to do so post-COVID. Warren discusses the roles and skills of chief information officers (CIOs), chief technology officers (CTOs) and enterprise architects to deliver value post-pandemic. Ultimately, success will follow those who can pivot quickly as the environment changes and those who can define a journey forward.

The concept of a roadmap to strategise for the future, something especially needed in time of crises, is at the centre of the work by Professor Agostino G. Bruzzone and Dr. Marina Massei of the University of Genoa. Bruzzone and Massei point out that many issues in current corporate strategies are due to a myopic time horizon that emphasize short term gains or objectives based on targets set on the next quarter or the next political campaign. When disruption does happen, not only do such approaches fail, but future recovery will be hindered. Nowadays, advances in crucial sectors of simulation, modelling, artificial intelligence and data analytics have a huge potential, especially when connected with their capabilities to process the massive data generated by companies and institutions as well as by technological advances, such as sensor networks and the Internet of

Things. All these new tools enable us to look at the future via the innovative lenses of 'strategic engineering', a new discipline merging technology and business insights to provide a closed loop approach to aid in the decision-making process.

The inherent risks of technology should not be overlooked, though, and these are thoroughly discussed in two chapters dedicated to the critical field of cyber security. In 'Cyber Security: Evolving Threats in an Ever-Changing World', Associate Professor Roberto Dillon of James Cook University joins a team of leading experts from KPMG Singapore, Paul Lothian, Simran Grewal and Daryl Pereira, to assess the current situation and the increased exposure of companies across all industries to a variety of renewed threats. Unfortunately, in the forced and sudden shift to remote working, security was often taken for granted with the consequence of opening up a digital risk gap where companies had to try to manage the scammers, fix their security infrastructures and launch new services securely, something they were often unprepared for. Attacks, such as stealing corporate credentials via phishing, have become widespread and a common occurrence. To provide a comprehensive picture of the new cyber security landscape we live in, the authors also explore how cyber security often gets intertwined with geo-politics and discuss high profile hacks like SolarWinds Orion, as well as widespread phishing techniques. All these examples are then brought together to exemplify how companies can reassess risk and start thinking about managing digital and cyber risks in the post-COVID world in a more efficient way.

In 'Reducing Cyber Risk in Remote Working', Dillon joins Vihangi Vagal, cyber security consultant for Deloitte UK, in an in-depth analysis dedicated to the now critical home-office environment and the additional risks that such set-up brought to every company. Spear phishing, ransomware attacks, exposure of sensitive information via social engineering, denial of service attacks, exploiting bugs in the virtual private network (VPN) and other such threat actors are terms remote employees need to familiarise themselves with while organisations devise new approaches to mitigate cyber risk and protect the confidentiality, integrity and availability of their resources.

While many sectors had huge difficulties in adapting, it is also important to understand how others did manage to embrace the social consequences of the pandemic and use them at their advantage as additional tailwind to fly towards new industry paradigms at an even greater pace. Professor Marco Accordi Rickards from the University Tor Vergata of Rome, joined by Micaela Romanini and Guglielmo De Gregori from the Vigamus Foundation, illustrates how and why online gaming has established itself as the most loved hobby during the different lockdowns. Rickards et al. argue that the changes set in motion by the new social restrictions have actually accelerated a shift towards cloud services that was already due to happen and that the flexibility of small to medium enterprises in this sector, unlike other entertainment industries, were very receptive to the changes.

PLANET

The different waves and reoccurrence of COVID-19 also meant unprecedented disruptions across a huge range of different global human activities. Some would even posit that such disruptions provided us with new opportunities not only for new cutting-edge technologies, like previously discussed in the first two sections of the book, but also for the planet as a way to rethink old procedures and practices that were proving ineffective or even harmful in the long term. The *Planet* section explores how we seized the moment to rethink alternative strategies towards the ways activities are organised and conducted, for example on education, city planning, and workplace transformation.

David J. Calkins in 'Digital Workplace Adaptation and the 20-Minute City' focuses the dramatic shift to the home office not from a technical perspective but from a more holistic one. The Gensler's Regional Managing Partner, in fact, discusses the dramatic challenges his global company had to face and considers whether it will be feasible to get back to the 'old ways' after the world is back to normal. It seems clear that, moving into the future, workers will expect to have the flexibility to continue to work from home. In the post-COVID world, the office will act as a facilitator for collaboration and social interaction. Companies may occupy less space as individual/focus work moves to homes or third places in the community. This trend may lead to significant changes in downtown areas, and landlords may begin to convert unused space to other uses to preserve their income streams. As a consequence, old commuting habits are going to be disrupted as people realise the time they have gained in lieu of commuting is a significant improvement to their lifestyle. This can have far reaching consequences, including the birth and development of new sub-urban areas, the so-called 20-Minute Cities, which may rise as important business centres in a not too far future.

Cities and landscapes are also at the centre of Adrienne Yuen's chapter, 'Nature, the Pandemic, and the Resilience of Cities: Case Study of Ottawa,

Canada'. Contrary to many countries where people were forced to stay at home in seemingly never-ending lockdowns, the COVID-19 pandemic has spurred many Canadians to spend more time outside, with an increase in outdoor activities such as running, hiking, biking, and use of public parks. In parallel, increased teleworking and time spent at home has driven a rise in demand for single-family homes, while the market for condominiums has softened. Taking the City of Ottawa as a case study, Yuen of the Standards Council of Canada explores these trends and outlines the consequent challenges and opportunities for sustainability. For example, we could have a potentially sustained trend of people moving out of the urban core and demanding larger homes with the accompanying need for land and infrastructure to service lower density communities. On the other hand, we also have an opportunity for creating more green spaces within cities, also called 're-greening', that could help maintain the attraction of urban living.

Modern, smart cities are also at the centre of rising privacy concerns, something that got into the spotlight recently as governments tried new ways to monitor citizens to track the pandemic. Melissa Wingard, a Senior Commercial Technology Lawyer at Phillips Ormonde Fizpatrick in Australia, reflects upon the disconnect commonly referred to as the 'privacy paradox', i.e. where people claim to care greatly about privacy, but then their actions rarely align with the concern expressed.

Among the many consequences of COVID-19, we saw the acceleration in the use of biometric data in technology to provide contactless solutions and implement contact tracing to address the unprecedented public health challenge. Biometric data is inherently identifiable as it uses our physical or behavioural human characteristics, such as our faces, eyes, gait, or finger prints, which are unique and unchangeable. Wingard highlights that as the information being collected moves from information about us to information that is fundamentally the essence of our being, the relative laws also need to be updated to empower individuals to manage and control their own biometric data so as to avoid being exploited.

The transformation of modern cities does not just affect areas such as landscaping or our privacy, though, but also includes other, far-reaching aspects of our lives, including education. Indeed, this sector was one that was hit the hardest as it had to reinvent itself overnight to accommodate new delivery methods. Professor Abhishek Bhati, together with Caroline Wong and Esther Fink of James Cook University, discusses these dramatic changes in affecting the education sector, too. By looking at a holistic approach to the on-boarding of students in the hybrid model of virtual with face-to-face experience, Bhati et al. examine how disruption can lead to innovation in how and when learners engage with digital content and activities to ensure that the learning outcomes are still achieved despite the disruption. The experiences of this sector, arguably, had to go back to the writing board and reimagine not only how to keep innovating, but also how to do old things in a novel innovative way.

PROSPERITY

In the *Prosperity* section, thought leaders like Esposito, Lanteri, Tse, Kuah and Lockie join forces to conclude with a discussion on the future of food, sustainable development goals, and the circular economy, including alternative thinking that may lead to a more sustainable future with digital transformation aiding in the process.

COVID-19 has acted as an accelerator for digital transformation and sustainability. Managers find themselves needing to re-evaluate and take steps to change on digital transformation. Decision makers may still face sudden black swans and wildcards that lead to unexpected circumstances and rapidly shifting competitive scenarios. Professors Mark Esposito, Alessandro Lanteri and Terence Tse from Hult International Business School and ESCP Business School have observed, pivoted, and researched several points of junction, where current trends set the trajectory for future events. In their chapter, Esposito, Lanteri and Tse provide their thought leadership on their macro-level DRIVE and meso-level CLEVER frameworks in empowering firms to identify growth opportunities in this turbulent, post-pandemic world economy.

Esposito et al. narrow their research down to five macro-level megatrends in the world, which they discuss in detail, post-pandemic: Demographic and social changes; resource scarcity; inequalities; volatility, scale and complexity; and, finally, enterprising dynamics. At the same time, they also share their framework of six strategic drivers: Collaborative intelligence; learning systems; exponential technologies; value facilitation; ethical championship; and responsive decision making that become ever-more important in the post-pandemic world.

The pandemic triggers considerations on how consumption and production patterns can be shifted to a more sustainable way. Sustainable consumption and production is about systemic change and 'doing more and better with less impact' to decouple economic growth from environmental degradation. Adrian T. H. Kuah, Associate Professor of Business, and Chang H. Kim of James Cook University argue how consumers have changed their consumption patterns as they transited to increased online entertainment, shopping, and food delivery. The growth of online shopping and 3D shopping, also noted by Tomczyk earlier, led to a surge in the generation of unnecessary packaging wastes. The pandemic has exacerbated the gravitas of the situation, and the new norm of online retail is likely to remain.

The core of circular economy models – resource recycle and recovery, remanufacturing, product life extension, sharing platforms, and product as a service – seek sustainability through optimisation of resources and away from one-time consumption. Echoing Mills' assertion that customers, investors and communities must transit into a new circular economy, Kuah and Kim propose incorporating digital technologies to a reverse logistics system to create new values in the circular loop. Companies can deliver more than ever values through the active introduction of digitised service strategies, such as smart remanufacturing and smart recycling. Digital servitisation aids to close the circular loop by further introducing servitisation alternatives, such as leasing, renting, and sharing. Hence, the total use of resources can be reduced and sustainable consumption maintained.

Furthering sustainable production and consumption, Stewart Lockie, Distinguished Professor of Sociology at the Cairns Institute, discusses barriers along with opportunities to develop physical, regulatory and institutional infrastructures that support socially and environmentally responsible innovation in agriculture and the future of food. Advanced sensing, automation, the Internet of Things, data analytics, artificial intelligence, synthetic biology, distributed ledger and other emerging technologies offer a seemingly endless array of possibilities to boost agricultural productivity. None of these outcomes, Lockie argues, are inevitable due to identifiable barriers to technological innovation and adoption, which must first be overcome to achieve sustainable production of food for prosperity.

As the old adage goes, 'where there are problems, there are opportunities' and this has never been more true than today: at a time of an unprecedented crisis in our lifetime, we all have the duty to increase our efforts, stepping up on the shoulders of the giants who preceded us and establish new paths to safely fare into uncharted territory that will not only make us survive, but also thrive. The aim of this volume then is to act as a 'brainstorming' tool across many different fields, where every contributor freely shared his or her expertise, experience and thoughts not only to outline the dramatic changes we are living in, but also, and most importantly, to inspire you, our reader, to seek new opportunities in the post-COVID world.

Adrian T. H. Kuah and Roberto Dillon
Singapore, 26 February 2021

NOTE

1. Newton, Isaac. "Letter from Sir Isaac Newton to Robert Hooke". Historical Society of Pennsylvania. Retrieved 7 June 2018.

I

People

The first section discusses the basis of human well-being, mental health, business and society that were heavily impacted by COVID-19, as well as how international political economy changes with trade tensions and restriction to travel affecting the future of business.

CHAPTER 1

COVID-19

Implications for Mental Health and Well-Being, Now and in the Digital Future

Caryn West[1] and Christopher Rouen[2]

[1]*College of Healthcare Sciences, James Cook University Singapore, Singapore*
[2]*College of Healthcare Sciences, James Cook University Singapore, Singapore*

CONTENTS

DOI: 10.1201/9781003148715-1

1.1 INTRODUCTION

The expeditious spread of the severe acute respiratory syndrome corona-virus 2 (COVID-19) has forced individuals and communities to rapidly adapt in the face of challenging and unprecedented circumstances. Named for the 'corona' or crown-like thorns on their surface, coronaviruses possess a distinct morphology and cause a broad range of diseases in both humans and animals (Tyrell, 1968). Seven coronaviruses are known to cause disease in humans, and of those, three are known to have resulted in severe to fatal respiratory infections (Tesini, 2020). The first of these, *SARS-CoV* was detected in China's Guangdong province towards the end of 2002 and presented as an influenza-like illness that, in a number of cases, progressed to severe respiratory insufficiency [severe acute respiratory syndrome (SARS)]. Between 2002 and 2004, *SARS-CoV* spread to over 30 countries with more than 8,000 cases reported worldwide (Institute of Medicine, 2004). The case fatality rate of *SARS-CoV* was approximately 10%, with individuals aged ≥ 65 years impacted most. Importantly, the *SARS-CoV* infectious outbreak was the first time that the Centers for Disease Control and Prevention (CDC) advised against travel to the regions affected (Institute of Medicine, 2004).

In September 2012, Middle East Respiratory Syndrome (MERS) caused by the MERS coronavirus (*MERS-CoV*) was reported in Saudi Arabia [World Health Organisation (WHO), 2018]. Causing mild to severe acute respiratory illness, approximately 2,500 cases of *MERS-CoV* have been reported from 27 countries to date. All transmissions have been linked through travel to, or residence in countries in and near the Arabian Peninsula; however, the largest outbreak occurred in the Republic of Korea during 2015 (Oh et al., 2018). Reflective of *SARS-CoV*, the median age of individuals affected by *MERS-CoV* was 54 years, with severe infection trends in patients with pre-existing conditions or comorbidities (WHO, 2018).

COVID-19 is an acute to severe respiratory illness caused by the novel coronavirus *SARS-CoV-2* (Esakandari et al., 2020). Believed to have originated in the Chinese city of Wuhan, *SARS-CoV-2* has spread to nearly every country on the planet and is considered by the CDC and the WHO to be a serious global health threat. Declared a pandemic on the 11 March 2020 (WHO, 2020), there were 106,865,939 confirmed COVID-19 cases and 2,338,004 deaths globally as of 10 February 2021 (Dong et al., 2020).

Like both *SARS-CoV* and *MERS-CoV*, the *SARS-CoV-2* virus has greatly impacted aged populations and individuals with existing conditions or comorbidities. What differs is the speed in which *SARS-CoV-2* is able to progress to acute respiratory distress syndrome, death and/or for seemingly mild and recovered cases, result in serious post-infectious complications (Cascella et al., 2020). Although it is known that coronaviruses can, will and do mutate, *SARS-CoV-2* has exhibited the ability to mutate quickly and the variant strains appear to be super virulent and exponentially difficult to contain (Abdullahi et al., 2020). At the time of writing, three particularly concerning variant strains of *SARS-COV-2* have been identified which have further complicated global pandemic response efforts (Public Health England, 2020, Wu et al., 2021, Faria et al., 2021).

Current research suggests that COVID-19 has resulted in an increase in the prevalence of psychological distress within general populations and an intensification of psychological distress among individuals with pre-existing mental health disorders. Additionally, the numerous uncertainties caused by COVID-19 may influence suicide rates and mental disorders associated with suicide in the foreseeable future. Despite these potential negative consequences on mental health, COVID-19 has been a catalyst for digital transformation in many sectors, including education, finance, business and healthcare. Globally, health systems have struggled to provide optimum service delivery and meet demand. Tertiary healthcare centres have seen systems and staff pushed beyond limits in a desperate attempt to respond to societal needs and as a result, have been forced to employ unique and innovative means of conducting day to day business. The dramatic upscaling of telehealth has facilitated the continued provision of healthcare services whilst minimising the risk of COVID-19 transmission potential.

In this chapter, we provide an outline of the impacts and implications of large-scale infectious disease outbreaks, why COVID-19 is considered the 'perfect storm' and why it has the potential to negatively impact mental health and well-being now and in the post-pandemic phase. We will examine the current evidence of the impacts on mental health and well-being and how the digital transformation of healthcare – lead by telehealth – can assist in defending and redefining mental health and mental healthcare.

1.2 THE UNPRECEDENTED PERFECT STORM

Throughout history, infectious disease outbreaks have resulted in large-scale demographic, economic and political disruption. However, none in recent history have had the crippling effects of COVID-19. Unlike other large scale infectious disease outbreaks, COVID-19 and the associated negative implications of well-intended and necessary prevention strategies have created a combination of widespread adverse factors, which have been described as an unprecedented 'perfect storm'. In recent history, there has been no equivalent infectious disease outbreak that has generated such widespread fear, drastic public health responses and considerable economic shock, that when combined, has the potential to significantly affect mental health and well-being on an unprecedented scale. The following explores in greater detail three major contributing elements that have been intrinsic in the creation of the COVID-19 'perfect storm'. Importantly, these elements cannot and should not be viewed only in relation to individuals who have known mental health diagnoses or are considered vulnerable. The uniqueness presented by COVID-19, particularly the perpetual cycle of the 'perfect storm' factors, has meant that vulnerability now exists where previously it did not.

1.2.1 Fear

Fear in itself is a natural inbuilt reaction that triggers biochemical, emotional and physical responses to perceived or real danger or threats. In most cases once the imminent threat has passed, the fear response subsides. However, fear is a multifaceted construct due to its subjectivity. Not all manifestations are homogenous nor all triggers universal. Some fears may be a result of experiences or trauma, while others may broadly represent a fear of the known or the unknown.

COVID-19 is unfamiliar, contagious, has invoked strong public health responses, changed everyday life, disrupted the social fabric and is associated with mortality. The very nature of a pandemic such as this justifiably creates fear. On a simplistic front, this fear allows individuals to confront and deal with potential threats, for example behaviours that assist with minimising the transmission of the virus, such as improving hand hygiene and encouraging adherence to physical distancing measures. However, depending on circumstances, personal and societal, the fear experienced may escalate resulting in a heightened and extended fear experience. It is this continued state of flux coupled with little or no respite that can result in negative mental health outcomes.

The difficulty COVID-19 poses for many is finding a 'fear' balance in a continuously changing landscape. Not enough fear creates the potential for conscious recklessness resulting in individual exposure to the virus and societal vulnerability due to breaches in public health measures. Excessive exhibitions of fear can result in anxiety, panic disorders, post-traumatic stress disorder (PTSD), social anxiety disorder and many other mental health conditions. At a societal level, excessive fear may result in fear induced irrational behaviours such as panic buying and xenophobia.

For health professionals, COVID-19 presents an interesting global mental health challenge as it has influenced every aspect of life and has created disruption to the social fabric. Fears that once were experienced by few, such as the uncertainty of a diagnosis and the wait for results, are now experienced on a grand scale, resulting in scepticism, distrust and intolerance where previously none existed.

1.2.2 Quarantine and Isolation

Quarantine, isolation and related public health responses have collectively resulted in diminished access to our typical social networks. Social support and social connectedness serve as important 'gatekeepers' to mental well-being (Lee et al., 2018) and for most, bolster necessary protective factors (Yoshioka-Maxwell, 2020). For known vulnerable or at-risk populations, the ability to engage socially and seek support is a well-documented mitigation strategy to deter risky behaviour (Tucker et al., 2015, Golden et al., 2009, Yoshioka-Maxwell and Rice, 2017). From an impact perspective, clear evidence exists regarding the negative impacts of social isolation on mental health, substance use/abuse, homelessness, interpersonal violence and risk behaviour (Tsai and Wilson, 2020, Usher et al., 2020, Perri et al., 2020). So it would stand to reason that in the event of a pandemic, social cohesion, support networks and interpersonal contact would be a critical means of maintaining both individual and community well-being and mitigating potential risk. When measures such as mandated quarantine and extended isolation interrupt those social connections, individuals are more likely to be exposed to factors that are associated with negative mental health outcomes that influence their health, safety and well-being.

1.2.3 Economic Shock

Across the globe, lives have been dramatically altered due to COVID-19. Global economic strain has stymied tourism, aviation, agriculture and finance, with all reporting drastic declines and massive reductions in both supply and demand (e.g. Ibn-Mohammed et al., 2021). Local and national economies have been crippled through sanctioned infection, prevention and control measures implemented with good intent, to reduce transmission rates (Nicola et al., 2020). In many countries, stimulus packages have been introduced in an effort to keep the economy afloat and stave off recession and potential financial collapse (Siddik, 2020). For the average person, this level of economic adversity increases stress levels and anxiety creating vulnerability, loss of identity, depression and loss of purpose (Frasquilho et al., 2016). Additional stress factors linked specifically to the *SARS-COV-2* pandemic are underemployment, job instability and school and childcare closures, none of which are counted in unemployment numbers (Douglas et al., 2020). For many single- and low-income families, increased financial demands associated with childcare, the loss of free school meals and having to heat the home during the day will push them over the solvency threshold and into the welfare system (Douglas et al., 2020). It is estimated that in the United Kingdom alone, 3.5 million additional individuals will need universal credit because of COVID-19 (Benstead, 2020).

A catalyst to the impacts of the global economic crisis created by COVOD-19 is the number of individuals employed in the 'gig economy'. This free market system, aimed to mobilise the workplace, is awash with temporary and independent workers employed for short-term commitments with no sick pay, vacation entitlements and often zero contract hours (Kuhn, 2016). As the pandemic continues, many within this group are at increased risk of rent or mortgage arrears and potential homelessness.

As with fear and isolation, there is a clearly established link between employment/income and health (Morris et al., 2000). Lower income earners are known to experience higher levels of psychological stress. When economies falter, it is often these workers that bear the brunt of the downturn with women and young workers often faring the worst (Douglas et al., 2020).

Within the health sector, the associated health consequences caused by economic demise are complicated. Gross unemployment negatively affects both physical and mental health exacerbating known diagnoses and increasing the risk of both homicides and suicide (Paul and Moser, 2009). A large-scale meta-analysis indicated a '76% increase in all-cause mortality in people followed for up to 10 years after becoming unemployed' (Roelfs et al., 2011). Looking forward into the post-pandemic landscape, these findings indicate there is a high likelihood of increased disadvantage and vulnerable and displaced individuals which could prohibit a hard reset of the employment sector leading to drawn out financial hardship.

As much of the world's population retreats indoors due to dropping temperatures, areas of congregate living, overcrowding, poor ventilation and high density populous calamitously combine to create the risk of yet another wave, resulting in further lockdowns and potential increased loss of life. Until a viable and globally accessible vaccine is available and widely implemented, governments around the world will continue to respond to the pandemic with a strong focus on public health measures aimed at eliminating, containing and/or slowing the virus. Although such strategies may well achieve the desired reduction in COVID-19 transmissions, the conditions of the unprecedented 'perfect storm' and potential cyclic nature of additional outbreaks will likely continue to have a significant impact on mental health and overall well-being.

1.3 COVID-19 AND MENTAL HEALTH

At the time of writing, the available evidence concerning the negative impact of COVID-19 on mental health is limited to the initial phase of the pandemic; although historic evidence from previous infectious disease outbreaks suggests that the consequences of COVID-19 are likely to be long-term. Despite this limitation and acknowledging the number of potential variables that could influence positive mental health outcomes, the current evidence remains damning and leads us to believe that the global impact on mental health will be considerable and linger after the pandemic has subsided. Individually, each element of the 'perfect storm' has the potential to cause widespread damage and poor mental health outcomes. Not only are the elements cumulative but they are ongoing and pervasive with little respite and no endpoint. The impact of COVID-19 on mental health will be influenced by the unique circumstances and experiences of the individual; however, for discussion, we examine the impacts on persons infected with COVID-19, healthcare workers and the general population.

1.3.1 Persons Who Have Been Infected with COVID-19

What would make people get it? … most adults have had a seriously unpleasant … illness. Maybe a persistent bad fever with hallucinations; a chest infection, with a painful cough, maybe with scary breathing problems; or severe systemic pain; vomiting and diarrhoea; dermatological problems; unusual bleeding; extreme weakness or fatigue; or something involving massive weight loss and muscle wasting; taste and smell disorders; frightening paraesthesia; joint pain. We've all had one or two of those things. They're in our lexicon of experience. On their own they're not alien. What is alien is having them all at once over many weeks. That's part of what severe COVID is: it's 5 or 10 illnesses at once.

Now throw into the mix that this is new and unknown, and that for most people there's no meaningful treatment or cure; that you might suddenly get worse, and not recover. You're facing death and there's nothing anyone can do. The other part of severe COVID … and my main abiding experience … that it's not in a hurry to leave my head.

(Stewart, 2020)

In a systematic review examining psychiatric and neuropsychiatric presentations associated with severe coronavirus infections (SARS or MERS), individuals admitted to hospital exhibited a number of common symptoms, including depressed mood, confusion, insomnia and impaired memory (Rogers et al., 2020). In the post-illness stage of these viruses, depressed mood, anxiety, irritability, insomnia, memory impairment and sleep disorder were also frequently reported. Point prevalence estimations suggested that in the post-illness stage, 32.2% of patients experienced PTSD, 14.8% experienced anxiety disorders and 14.9% experienced depression (Rogers et al., 2020). For individuals who also had known mental health diagnoses, aggravated expressions of symptomologies were witnessed (Jeong et al., 2016).

In individuals admitted with acute COVID-19 symptomology to isolation wards in Wuhan, China, depression and anxiety were reported to have a prevalence of 28–34%, (Kong et al., 2020). Similar data collected from clinically stable patients with COVID-19 in Hubei Provence, China reported a higher depression prevalence of 43.1% (Ma et al., 2020). Data collected from recently recovered COVID-19 patients in Guangdong Province, China reported a depression prevalence of 30% (Zhang et al., 2020). Although the sample sizes presented are relatively small, the evidence suggests that on average, regardless of the phase of illness (acute, stable or recovering) there is potentially a 1 in 3 chance of an individual experiencing some form of depression or anxiety due to COVID-19. Supporting these findings, prevalence estimates of up to 96.2% for post-traumatic stress symptoms have been reported in stable COVID-19 patients (Bo et al., 2020) and indicative evidence exists to support the development of post-viral syndromes with similar symptomology to depression (Troyer et al., 2020, Lyons et al., 2020).

Applied in isolation these findings do not seem overly impactful; however, given the number of COVID-19 infections identified globally, the mental health burden associated with individuals infected with COVID-19 will be orders of magnitude greater than what was globally evidenced by both SARS and MERS.

1.3.2 Healthcare Workers

> To YOU, the healthcare workers but also the human beings behind the PPE, we want to encourage you to continue, ... you are in our thoughts, in our hearts and prayers... "take it one day at the time", "don't look in the rear view mirror anymore but only forward"..."don't let scary thoughts win, continue the fight and NEVER quit"... YOU WILL BEAT THE VIRUS, the fatigue and the discouragement because the STRENGTH and THE WORLD is within you!
>
> *(Walker, 2020)*

Infectious disease outbreaks place healthcare workers in high-stress environments. During the SARS epidemic, exposed healthcare workers in Taiwan who were quarantined reported symptoms of acute stress disorder, higher levels of exhaustion, detachment, irritability, insomnia and anxiety (Bai et al., 2004). Similarly, a Chinese study assessing the impact of SARS on hospital healthcare workers reported that workers who were quarantined or worked in high exposure locations were two to three times more likely to have high post-traumatic stress levels when compared to individuals who did not quarantine or had lower exposure risk (Wu et al., 2009). Importantly, in 40% of respondents, symptomology persisted for longer than three years, highlighting the lingering impact such events can have on mental health and well-being.

The increasing pressure applied to health systems because of COVID-19 has accordingly placed significant mental health challenges on healthcare workers. Exacerbated by multiple and at times conflicting factors, challenges include: increased risk of COVID-19 infection, rising death rates, frustration, exhaustion, discrimination, media hype, isolation from social networks and supports, viral contamination of self, family and friends, moral injury and inadequate support (Lai et al., 2020, Dos Santos et al., 2020). A review of 11 studies investigating the mental health of healthcare workers in the early stages of the pandemic (until April 2020) reported that 24.1–67% of healthcare workers experienced anxiety, 12.1–55.89% experienced depression and 29.8–62.99% experienced stress (Vizheh et al., 2020).

More severe symptoms were reported in nurses, females, front-line health-care workers, younger medical staff and workers in areas with higher infection rates. This is particularly concerning given nursing accounts for nearly 50% of the global health workforce and is a predominantly female profession (WHO, 2016). Global data published until May 2020 indicated that 71.6% of COVID-19 infections experienced by healthcare workers occurred in women and 38.6% of infections occurred in nurses (Bandyopadhyay et al., 2020). In a pandemic that as yet, has no foresee-able end, coupled with increasingly virulent strains, the negative mental health impacts experienced by healthcare workers could have significant consequences beyond those associated with individual mental health deterioration.

Despite the numerous concomitant factors associated with COVID-19, by its simplest definition it is a global health emergency. It has shown that it does not discriminate, and that the very workforce charged with con-tainment and cure are just as susceptible, if not more, than the average person. As global health workforces continue to buckle under the strain of COVID-19 considerable support needs to be provided for the healthcare workers, now and in the post-pandemic phase.

1.3.3 General Population

At a population level, the impact of COVID-19 is likely to manifest uniquely and negatively influence mental health based on a variety of conditions and experiences relating to each individual's circumstances. In a system-atic review on COVID-19 impacts on mental health in general popula-tions across China, Spain, Italy, Iran, the United States, Turkey, Nepal and Denmark, results showed high prevalence of anxiety (6.33–50.9%), depres-sion (14.6–48.3%), PTSD (7–53.8%), psychological distress (24.43–38%) and stress (8.1–81.9%) (Xiong et al., 2020). Additionally, females aged ≤ 40 years, individuals with chronic/psychiatric illnesses, unemployment, student status, and frequent exposure to social media/news concerning COVID-19 also reported high distress measures.

Physical distancing, quarantine requirements and/or fear have also resulted in reported impacts on mental health. Documented evidence includes: phobic anxiety (Tian et al., 2020), panic buying (Sim et al., 2020); and an increase in binge watching habits, leading to mood disturbances, sleep disturbances, fatigability and impairment of self-regulation (Dixit et al., 2020). Increased exposure to social media has similarly been asso-ciated with amplified anxiety and anxiety combined with depression

(Gao et al., 2020). Isolated research suggests that individuals placed in mandatory quarantine have experienced a fivefold increase in the likelihood of self-harm or suicidal ideation (Xin et al., 2020). Importantly, these results are not necessarily generalisable to global populations, and attempts to homogenise data sets could lead to misrepresentation of the mental health impact and associated risk factors.

As with previous sections, the burden of quarantine, isolation or lockdown attributed to vulnerable populations is disproportionate when compared to the general populous. Preliminary evidence suggests aggravation of symptoms for anxiety, depression, PTSD, insomnia (Hao et al., 2020) and significant socioeconomic disadvantage. Confinement to home, disruption of daily routines and physical distancing not only exacerbate conditions but present ongoing challenges for vulnerable individuals as well as caregivers and healthcare providers (Yao et al., 2020, Hao et al., 2020).

1.3.4 Into the Future

It is clear that concerted efforts are being employed to scaffold the broader mental health challenges and the compounding effects of the pandemic. However, continued uncertainty and fear associated with COVID-19, combined with drastic public health responses and economic shock raises significant concerns for long-term mental health and well-being. If these concerns are not effectively addressed with robust and long-term intervention strategies, the mental health burden associated with the pandemic may persist well into the future, including potential increases in self-harm and suicide.

At a population level, suicide rates are highly sensitive to macroeconomic indicators, principally unemployment. (Chang et al., 2009, Reeves et al., 2012, Stuckler et al., 2009). The 2007–2008 Global Financial Crisis, saw increased unemployment with a resultant 10,000 additional economic suicides across North America and Europe (Reeves et al., 2012). Similarly, evidence from the SARS epidemic was also associated with an increase in deaths by suicide (Zortea et al., 2020). Given the strong association between suicide and mental health disorders and the current economic downturn and collapse of the global job sector, precedence suggests this is an area of urgent need.

While the current focus of governments and healthcare providers is on effective minimisation of impacts due to COVID-19, determining the long-term effects on mental health is a challenge that needs to be addressed proactively rather than reactively. Much of the mental health burden associated with COVID-19 is either preventable or treatable through early intervention and access to health services. In the current climate, this will mean a seismic shift in not only health service delivery, but how health delivery and access is perceived by the general public.

1.4 THE DIGITAL TRANSFORMATION OF MENTAL HEALTH SERVICE PROVISION

Historically, healthcare systems have been structured upon in-person interactions between providers and consumers. This has inherently resulted in the congregation of consumers into central health service delivery hubs, such as emergency departments and primary healthcare waiting rooms, particularly in times of crisis. Infectious disease outbreaks therefore pose significant challenges in ensuring the provision of ongoing healthcare services whilst simultaneously addressing outbreak-related service demand and the minimisation of transmission risk. In the context of COVID-19, these challenges are further compounded by physical distancing requirements; mass quarantine of large geographic areas; redeployment of the healthcare workforce to address COVID-19; repurposing of existing services and facilities to the pandemic response; and inadequate supply of personal protective equipment. COVID-19 has therefore effectively lowered the capacity for health systems to respond to the demand for healthcare. This has resulted in difficulties in accessing specialty health services across the health spectrum (Iob et al., 2020).

1.4.1 The Digital Transformation of Telehealth

The unique circumstances presented by COVID-19 have catalysed the rapid uptake of telehealth which has facilitated the provision of healthcare services whilst also protecting providers, consumers and the broader community from exposure to the virus (Shore et al., 2020). Broadly, telehealth encompasses a range of services and activities performed at distance that incorporates provider-to-consumer and/or provider-to-provider interactions (Wosik et al., 2020). Telehealth can occur synchronously through video or telephone conferencing; asynchronously through email or messaging portals; or via virtual agents, such as chat bots (Hills and Hills, 2019, Ting et al., 2020). Although telehealth has long been touted as an effective means to provide healthcare, its implementation has been slow and fragmented due to a variety of challenges associated with regulators, service providers and consumers. However, as the standard model of in-person care became unviable in the early stages of the pandemic, and government policy shifted to support telehealth, its utilisation in healthcare systems globally has surged (Moore and Munroe, 2020).

A previous review examining the merit of telehealth interventions for mental health conditions identified strong and consistent evidence that suggested telehealth interventions were a feasible model of care, acceptable to consumers, improved symptoms, improved quality of life and resulted in positive cost savings (Bashshur et al., 2016). This evidence base includes mental health conditions that appear to have increased in prevalence during the COVID-19 pandemic, such as depression (García-Lizana and Muñoz-Mayorga, 2010), anxiety (Rees and Maclaine, 2015) and post-traumatic stress symptoms (Turgoose et al., 2017). In addition, telehealth has shown clear advantages in the context of COVID-19 including: (1) improved access to mental health services for the general population; (2) a reduction in resource utilisation within healthcare facilities; (3) a mechanism to maintain continuity of care for consumers with a pre-existing or newly diagnosed mental health disorders; (4) facilitation of more frequent provider-consumer engagements; (5) improved access in healthcare systems with traditionally low investment and low capacity for mental health services; (6) creation of risk mitigation strategies for vulnerable populations during the pandemic and in future outbreaks and (7) enabling early intervention to prevent escalation of existing conditions, such as the detection of consumers at risk of self-harm and hospitalisation prevention (Galea et al., 2020, Chew et al., 2020, Tan et al., 2015, Killgore et al., 2020, Naslund et al., 2017, Torous and Wykes, 2020).

1.4.2 The Challenges of Telehealth

Despite these clear advantages, there are however, considerable challenges that need to be overcome to support this digital transformation for the remainder of the pandemic and to ensure that telehealth for mental health services is sustainable in a post-pandemic world (Smith et al., 2020). These include: (1) appropriate and sustainable remuneration for telehealth services; (2) knowledge and training for providers and consumers; (3) suitable internet access infrastructure; (4) potential exacerbation of the digital divide and (5) addressing privacy and data security issues (Smith et al., 2020). Furthermore, in some populations, such as the elderly, those with limited digital literacy skills and the socioeconomically disadvantaged, may find telehealth more challenging to access, which highlights the need to maintain some form of in-person service where telehealth proves to be inappropriate.

1.4.3 Digital Transformation into the Future

Importantly, telehealth is only one pillar of a larger digital transformation occurring in healthcare that has been mobilised by the COVID-19 pandemic. Big data systems are processing vast quantities of travel data and health system capacity to identify potential outbreak sites; online health communities are attempting to address the spread of viral misinformation; and chat bots are being developed to provide screening services at healthcare facilities (Bogoch et al., 2020, Figueroa and Aguilera, 2020, Espinoza et al., 2020). Such technologies are inextricably linked and must be developed and deployed collectively, guided by the needs of the consumers, the acceptability of the technology to providers and acceptability to regulators to ensure mental healthcare needs during and after the pandemic are met. To realise the long-term benefits of this transformation, healthcare providers, consumers and other stakeholders must collaborate to understand what worked well during the COVID-19 pandemic, what didn't, what should change, and why. Ongoing support is critical to ensure digital health technologies in the post pandemic world address the current challenges identified, and shift from ad hock systems developed to address need in a time of crisis to secure and suitable systems that preserve data security and patient privacy in a post crisis world.

1.5 CONCLUSION

COVID-19 has facilitated a 'perfect storm' of adverse factors that could potentially lead to a significantly universal mental health burden both during the pandemic and into the future. Fear, necessary public health measures and economic shock will have widespread mental health consequences on a truly global scale. The expeditious spread of the virus has necessarily catalysed a digital transformation in health provision, particularly telehealth, to obviate the need for physical interactions between healthcare providers and consumers. This digital transformation provides a means to improve access to mental health services, address the mental health concerns of the pandemic, provide support in a post-pandemic world and importantly, help prepare for future outbreaks. However, in order for telehealth to effectively address the negative mental health impacts of COVID-19, it will require continued support to transform it from reactive crisis response into a long-term and sustainable model of care.

REFERENCES

Abdullahi, I. N., Emeribe, A. U., Ajayi, O. A., Oderinde, B. S., Amadu, D. O., et al. 2020. Implications of SARS-CoV-2 genetic diversity and mutations on pathogenicity of the COVID-19 and biomedical interventions. *J Taibah Univ Med Sci*, 15, 258–264.

Bai, Y., Lin, C. C., Lin, C. Y., Chen, J. Y., Chue, C. M., et al. 2004. Survey of stress reactions among health care workers involved with the SARS outbreak. *Psychiatr Serv*, 55, 1055–1057.

Bandyopadhyay, S., Baticulon, R. E., Kadhum, M., Alser, M., Ojuka, D. K., et al. 2020. Infection and mortality of healthcare workers worldwide from COVID-19: a systematic review. *BMJ Global Health*, 5, e003097.

Bashshur, R. L., Shannon, G. W., Bashshur, N. & Yellowlees, P. M. 2016. The Empirical Evidence for Telemedicine Interventions in Mental Disorders. *Telemed J E Health*, 22, 87–113.

Benstead, S. 2020. Coronavirus to force 3.5 million extra people on to universal credit. *The Telegraph*.

Bo, H. X., Li, W., Yang, Y., Wang, Y., Zhang, Q., et al. 2020. Posttraumatic stress symptoms and attitude toward crisis mental health services among clinically stable patients with COVID-19 in China. *Psychol Med*, 1–2.

Bogoch, I. I., Watts, A., Thomas-Bachli, A., Huber, C., Kraemer, M. U. G., et al. 2020. Pneumonia of unknown aetiology in Wuhan, China: potential for international spread via commercial air travel. *Journal of Travel Medicine*, 27.

Cascella, M., Rajnik, M., Cuomo, A., Dulebohn, S. C. & Di Napoli, R. 2020. Features, Evaluation, and Treatment of Coronavirus. [Updated 2020 Oct 4]. *StatPearls [Internet]*. Treasure Island (FL): StatPearls.

Chang, S. S., Gunnell, D., Sterne, J. A., Lu, T. H. & Cheng, A. T. 2009. Was the economic crisis 1997–1998 responsible for rising suicide rates in East/Southeast Asia? A time-trend analysis for Japan, Hong Kong, South Korea, Taiwan, Singapore and Thailand. *Soc Sci Med*, 68, 1322–31.

Chew, A. M. K., Ong, R., Lei, H.-H., Rajendram, M., K V, G., et al. 2020. Digital Health Solutions for Mental Health Disorders During COVID-19. *Front Psychiatry*, 11, 898.

Dixit, A., Marthoenis, M., Arafat, S. M. Y., Sharma, P. & Kar, S. K. 2020. Binge watching behavior during COVID 19 pandemic: A cross-sectional, cross-national online survey. *Psychiatry Res*, 289, 113089.

Dong, E., Du, H. & Gardner, L. 2020. An interactive web-based dashboard to track COVID-19 in real time. *Lancet Infect Dis*, 20, 533–34.

Dos Santos, C. F., Picó-Pérez, M. & Morgado, P. 2020. COVID-19 and Mental Health-What Do We Know So Far? *Front Psychiatry*, 11, 565698.

Douglas, M., Katikireddi, S. V., Taulbut, M., Mckee, M. & Mccartney, G. 2020. Mitigating the wider health effects of covid-19 pandemic response. *BMJ*, 369, m1557.

Esakandari, H., Nabi-Afjadi, M., Fakkari-Afjadi, J., Farahmandian, N., Miresmaeili, S.-M., et al. 2020. A comprehensive review of COVID-19 characteristics. *Biol Proced Online*, 22, 19.

Espinoza, J., Crown, K. & Kulkarni, O. 2020. A guide to chatbots for COVID-19 screening at pediatric healthcare facilities. *JMIR Public Health Surveill*, 6, e18808–e18808.

Faria, N. R., Claro, I. M., Candido, D., Moyses Franco, L. A., Andrade, P. S., et al. 2021. *Genomic characterisation of an emergent SARS-CoV-2 lineage in Manaus: preliminary findings* [Online]. Available: https://virological.org/t/genomic-characterisation-of-an-emergent-sars-cov-2-lineage-in-manaus-preliminary-findings/586 [Accessed January 2021].

Figueroa, C. A. & Aguilera, A. 2020. The need for a mental health technology revolution in the COVID-19 pandemic. *Front Psychiatry*, 11, 523.

Frasquilho, D., Matos, M. G., Salonna, F., Guerreiro, D., Storti, C. C., et al. 2016. Mental health outcomes in times of economic recession: a systematic literature review. *BMC Public Health*, 16, 115.

Galea, S., Merchant, R. M. & Lurie, N. 2020. The mental health consequences of COVID-19 and physical distancing: The need for prevention and early intervention. *JAMA Internal Med*, 180, 817818.

Gao, J., Zheng, P., Jia, Y., Chen, H., Mao, Y., et al. 2020. Mental health problems and social media exposure during COVID-19 outbreak. *PLoS One*, 15, e0231924.

García-Lizana, F. & Muñoz-Mayorga, I. 2010. Telemedicine for depression: a systematic review. *Perspect Psychiatr Care*, 46, 119–26.

Golden, J., Conroy, R. M., Bruce, I., Denihan, A., Greene, E., et al. 2009. Loneliness, social support networks, mood and wellbeing in community-dwelling elderly. *Int J Geriatr Psychiatry*, 24, 694–700.

Hao, F., Tan, W., Jiang, L., Zhang, L., Zhao, X., et al. 2020. Do psychiatric patients experience more psychiatric symptoms during COVID-19 pandemic and lockdown? A case-control study with service and research implications for immunopsychiatry. *Brain Behav Immun*, 87, 100–06.

Hills, W. & Hills, K. 2019. Virtual treatment in an integrated primary care-behavioral health practice: An overview of synchronous telehealth services to address rural urban disparities in mental health care. *Med Sci Pulse*, 13, 54–59.

Ibn-Mohammed, T., Mustapha, K. B., Godsell, J., Adamu, Z., Babatunde, K. A., et al. 2021. A critical analysis of the impacts of COVID-19 on the global economy and ecosystems and opportunities for circular economy strategies. *Resour Conserv Recycl*, 164, 105169.

Institute of Medicine. 2004. *Learning from SARS: Preparing for the Next Disease Outbreak: Workshop Summary*. Washington, DC, The National Academies Press.

Iob, E., Steptoe, A. & Fancourt, D. 2020. Abuse, self-harm and suicidal ideation in the UK during the COVID-19 pandemic. *Br J Psychiatry: Jo Mental Sci*, 217, 543–546.

Jeong, H., Yim, H. W., Song, Y. J., Ki, M., Min, J. A., et al. 2016. Mental health status of people isolated due to Middle East Respiratory Syndrome. *Epidemiol Health*, 38, e2016048.

Killgore, W. D. S., Cloonan, S. A., Taylor, E. C. & Dailey, N. S. 2020. Loneliness: A signature mental health concern in the era of COVID-19. *Psychiatry Res*, 290, 113117.

Kong, X., Zheng, K., Tang, M., Kong, F., Zhou, J., et al. 2020. Prevalence and factors associated with depression and anxiety of hospitalized patients with COVID-19. *MedRxiv*.

Kuhn, K. M. 2016. The rise of the "gig economy" and implications for understanding work and workers. *Indust Organiz Psychol*, 9, 157–162.

Lai, J., Ma, S., Wang, Y., Cai, Z., Hu, J., et al. 2020. Factors associated with mental health outcomes among health care workers exposed to coronavirus disease 2019. *JAMA Netw Open*, 3, e203976.

Lee, S., Chung, J. E. & Park, N. 2018. Network environments and well-being: An examination of personal network structure, social capital, and perceived social support. *Health Commun*, 33, 22–31.

Lyons, D., Frampton, M., Naqvi, S., Donohoe, D., Adams, G., et al. 2020. Fallout from the COVID-19 pandemic – should we prepare for a tsunami of post viral depression? *Ir J Psychol Med*, 37, 295–300.

Ma, Y. F., Li, W., Deng, H. B., Wang, L., Wang, Y., et al. 2020. Prevalence of depression and its association with quality of life in clinically stable patients with COVID-19. *J Affect Disord*, 275, 145–148.

Moore, M. A. & Munroe, D. D. 2020. COVID-19 brings about rapid changes in the telehealth landscape. *Telemedicine and e-Health*.

Morris, J. N., Donkin, A. J. M., Wonderling, D., Wilkinson, P. & Dowler, E. A. 2000. A minimum income for healthy living. *J Epidemiol Commun Health*, 54, 885–889.

Naslund, J. A., Aschbrenner, K. A., Araya, R., Marsch, L. A., Unützer, J., et al. 2017. Digital technology for treating and preventing mental disorders in low-income and middle-income countries: A narrative review of the literature. *Lancet Psychiatry*, 4, 486–500.

Nicola, M., Alsafi, Z., Sohrabi, C., Kerwan, A., Al-Jabir, A., et al. 2020. The socioeconomic implications of the coronavirus pandemic (COVID-19): A review. *Int J Surg*, 78, 185–193.

Oh, M.-D., Park, W. B., Park, S.-W., Choe, P. G., Bang, J. H., et al. 2018. Middle East respiratory syndrome: what we learned from the 2015 outbreak in the Republic of Korea. *Korean J Internal Med*, 33, 233–246.

Paul, K. I. & Moser, K. 2009. Unemployment impairs mental health: Meta-analyses. *J Vocation Behav*, 74, 264–82.

Perri, M., Dosani, N. & Hwang, S. W. 2020. COVID-19 and people experiencing homelessness: challenges and mitigation strategies. *Can Med Assoc J*, 192, E716.

Public Health England. 2020. Public Health England. Investigation of novel SARS-CoV-2 variant: Variant of Concern 202012/01. Technical briefing 3.

Rees, C. S. & Maclaine, E. 2015. A systematic review of videoconference-delivered psychological treatment for anxiety disorders. *Aust Psychol*, 50, 259–64.

Reeves, A., Stuckler, D., Mckee, M., Gunnell, D., Chang, S. S., et al. 2012. Increase in state suicide rates in the USA during economic recession. *Lancet*, 380, 1813–14.

Roelfs, D. J., Shor, E., Davidson, K. W. & Schwartz, J. E. 2011. Losing life and livelihood: A systematic review and meta-analysis of unemployment and all-cause mortality. *Soc Sci Med*, 72, 840–54.

Rogers, J. P., Chesney, E., Oliver, D., Pollak, T. A., Mcguire, P., et al. 2020. Psychiatric and neuropsychiatric presentations associated with severe coronavirus infections: A systematic review and meta-analysis with comparison to the COVID-19 pandemic. *Lancet Psychiatry*, 7, 611–27.

Shore, J. H., Schneck, C. D. & Mishkind, M. C. 2020. Telepsychiatry and the coronavirus disease 2019 pandemic—current and future outcomes of the rapid virtualization of psychiatric care. *JAMA Psychiatry*, 77, 1211–12.

Siddik, M. N. A. 2020. Economic stimulus for COVID-19 pandemic and its determinants: Evidence from cross-country analysis. *Heliyon*, 6, e05634–e05634.

Sim, K., Chua, H. C., Vieta, E. & Fernandez, G. 2020. The anatomy of panic buying related to the current COVID-19 pandemic. *Psychiatry Res*, 288, 113015.

Smith, A. C., Thomas, E., Snoswell, C. L., Haydon, H., Mehrotra, A., et al. 2020. Telehealth for global emergencies: Implications for coronavirus disease 2019 (COVID-19). *J Telemed Telecare*, 26, 309–313.

Stewart, G. 2020. My experience of COVID-19 – not just another experience. Available from: https://blogs.bmj.com/bmj/2020/07/20/grant-stewart-my-experience-of-covid-19-not-just-another-experience/ [Accessed February 2021].

Stuckler, D., Basu, S., Suhrcke, M., Coutts, A. & Mckee, M. 2009. The public health effect of economic crises and alternative policy responses in Europe: An empirical analysis. *Lancet*, 374, 315–23.

Tan, S. C., Lee, M. W., Lim, G. T., Leong, J. J. & Lee, C. 2015. Motivational interviewing approach used by a community mental health team. *J Psychosoc Nurs Ment Health Serv*, 53, 28–37.

Tesini, B. L. 2020. *Coronaviruses and Acute Respiratory Syndromes (COVID-19, MERS, and SARS)* [Online]. MSD Manual – Professional Version. Available: https://www.msdmanuals.com/professional/infectious-diseases/respiratory-viruses/coronaviruses-and-acute-respiratory-syndromes-covid-19-mers-and-sars [Accessed February 2021].

Tian, F., Li, H., Tian, S., Yang, J., Shao, J., et al. 2020. Psychological symptoms of ordinary Chinese citizens based on SCL-90 during the level I emergency response to COVID-19. *Psychiatry Res*, 288, 112992.

Ting, D. S., Gunasekeran, D. V., Wickham, L. & Wong, T. Y. 2020. Next generation telemedicine platforms to screen and triage. *Br J Ophthalmol*, 104, 299–300.

Torous, J. & Wykes, T. 2020. Opportunities from the coronavirus disease 2019 pandemic for transforming psychiatric care with telehealth. *JAMA Psychiatry*, 77, 1205–1206.

Troyer, E. A., Kohn, J. N. & Hong, S. 2020. Are we facing a crashing wave of neuropsychiatric sequelae of COVID-19? Neuropsychiatric symptoms and potential immunologic mechanisms. *Brain Behav Immun*, 87, 34–39.

Tsai, J. & Wilson, M. 2020. COVID-19: A potential public health problem for homeless populations. *Lancet Public Health*, 5, e186–e187.

Tucker, J. A., Cheong, J., Chandler, S. D., Crawford, S. M. & Simpson, C. A. 2015. Social networks and substance use among at-risk emerging adults living in disadvantaged urban areas in the southern United States: A cross-sectional naturalistic study. *Addiction*, 110, 1524–32.

Turgoose, D., Ashwick, R. & Murphy, D. 2017. Systematic review of lessons learned from delivering tele-therapy to veterans with post-traumatic stress disorder. *J Telemed Telecare*, 24, 575–585.

Tyrell, D. 1968. Coronaviruses. *Nature (London)*, 220, 650.

Usher, K., Bhullar, N. & Jackson, D. 2020. Life in the pandemic: Social isolation and mental health. *J Clin Nurs*, 29, 2756–2757.

Vizheh, M., Qorbani, M., Arzaghi, S. M., Muhidin, S., Javanmard, Z., et al. 2020. The mental health of healthcare workers in the COVID-19 pandemic: A systematic review. *J Diabetes Metabolic Disord*, 19, 1967–1978.

Walker, A. 2020. This Compilation of Kind Words Written By Strangers, To Nurses, Will Make You Smile. *Nurse.org*.

World Health Organisation. 2016. Global strategic directions for strengthening nursing and midwifery 2016–2020. Geneva.

World Health Organisation. 2018. *WHO MERS-CoV global summary and assessment of risk—August 2018* [Online]. Available: www.who.int/csr/disease/coronavirus_infections/risk-assessment-august-2018.pdf [Accessed January 2021].

World Health Organisation. 2020. WHO Director-General's opening remarks at the media briefing on COVID-19-11 March 2020. Geneva, Switzerland.

Wosik, J., Fudim, M., Cameron, B., Gellad, Z. F., Cho, A., et al. 2020. Telehealth transformation: COVID-19 and the rise of virtual care. *J Am Med Inform Assoc*, 27, 957–62.

Wu, K., Werner, A. P., Moliva, J. I., Koch, M., Choi, A., et al. 2021. mRNA-1273 vaccine induces neutralizing antibodies against spike mutants from global SARS-CoV-2 variants. *bioRxiv*, 2021.01.25.427948.

Wu, P., Fang, Y., Guan, Z., Fan, B., Kong, J., et al. 2009. The psychological impact of the SARS epidemic on hospital employees in China: exposure, risk perception, and altruistic acceptance of risk. *Can J Psychiatry. Revue canadienne de psychiatrie*, 54, 302–311.

Xin, M., Luo, S., She, R., Yu, Y., Li, L., et al. 2020. Negative cognitive and psychological correlates of mandatory quarantine during the initial COVID-19 outbreak in China. *Am Psychol*, 75, 607–17.

Xiong, J., Lipsitz, O., Nasri, F., Lui, L. M. W., Gill, H., et al. 2020. Impact of COVID-19 pandemic on mental health in the general population: A systematic review. *J Affective Disord*, 277, 55–64.

Yao, H., Chen, J.-H. & Xu, Y.-F. 2020. Patients with mental health disorders in the COVID-19 epidemic. *Lancet Psychiatry*, 7, e21.

Yoshioka-Maxwell, A. 2020. Social Work in Action: Social Connectedness and Homelessness Amidst a Pandemic: Are the Social Impacts of Quarantine on Homeless Populations Being Adequately Addressed? *Hawai'i J Health Social Welfare*, 79, 329–331.

Yoshioka-Maxwell, A. & Rice, E. 2017. Exploring the impact of network characteristics on substance use outcomes among homeless former foster youth. *Int J Public Health*, 62, 371–378.

Zhang, J., Lu, H., Zeng, H., Zhang, S., Du, Q., et al. 2020. The differential psychological distress of populations affected by the COVID-19 pandemic. *Brain, Behav Immunity.*

Zortea, T. C., Brenna, C. T. A., Joyce, M., Mcclelland, H., Tippett, M., et al. 2020. The impact of infectious disease-related public health emergencies on suicide, suicidal behavior, and suicidal thoughts. *Crisis*, 1–14.

CHAPTER 2

On Leadership

Victor Mills

Singapore International Chamber of Commerce, Singapore

Leadership matters in business and on the national stage. There has never been a greater need for more decent men and women to lead companies and countries than there is today. Our world is beset by so many challenges, many of them existential like the pandemic and climate change. These challenges, in turn, hamper economic activity and growth, increase poverty, inequality and mass migration. And yet, when I look around the world today, I see a scarcity of decent leadership in action.

There is far too much selfishness in those leading companies and countries resulting in authoritarianism, bombast, division, dysfunction, hatred and toxicity. What the world needs now is collaborative leadership to deal with the challenges which affect all of us humans and our small planet.

If the COVID-19 pandemic has done anything it has increased peoples' appreciation of authenticity, of the ability to communicate with empathy and concern and to act decisively in the best interests of the whole community. Jacinda Ardern, the recently re-elected Prime Minister of New Zealand, personifies these positive traits and the electorate rewarded her for them. The pandemic has also highlighted what happens when leaders do the opposite of Ms. Ardern: people in large numbers die unnecessarily.

Selfless leadership is not only required by corporate bosses and country leaders. It is required by and from each of us. It is what I call self-leadership. The ability to maintain control, a sense of fairness, good humour,

DOI: 10.1201/9781003148715-2

reasonableness and the willingness to treat others as we would like to be treated. That is with respect. There is no greater form of respect these days than being able to put yourself in someone else's shoes and see the world from their perspectives provided all parties can agree on the facts of the issue.

Is this asking too much? I do not think so. Simply put, we have no choice but to demand these qualities from our corporate and political leaders and from ourselves if we are to stand any chance of providing a future for our planet and for succeeding generations of humans.

Leadership starts with each of us caring about ourselves and our communities. It is always we and never just me. Rights always come with limits and with responsibilities for us and for others.

The notion of individualism can, when carried to the extreme, breed the notion that no-one else matters. Nothing could be further from the truth. We are all individuals who live in communities. No man or woman is an island. The pandemic has shown that none of us are safe until we are all safe. Same goes for climate change.

It is not possible to achieve consistent success in business or in life without trust. Trust – in people, in processes, in institutions, in brands and in governance – is the essential enabling foundation on which relationships and collaborations are built and sustained. Trust must be built by each of us as individual leaders as well as by those who run companies and countries. Trust must be earned and sustained. Why it is in such short supply globally and what can we do to change that?

Trust is in such short supply partly because trust is hard to build. It takes consistent effort. You have to care enough about others enough to want to build trust. Conversely, it is easy to undermine and to lose. Trust is in short supply partly because of bad behaviour in business and in public life which is often criminal in nature. Trust is in short supply when people feel excluded and believe they have no stake in a community, system or control over outcomes which affect their lives. Trust is in short supply when institutions are pickled in aspic and no longer fit for purpose and where reform has stalled or is not even contemplated. When trust is eroded the weeds of cynicism and contempt flourish. The jungle grows back.

The development of social media has turbo-charged the undermining of trust across the board. It is an unintended consequence of tools which can and do achieve positive outcomes like connecting long lost relatives and friends or crowdsourcing to help fund someone's cancer treatment.

It is new media and insufficiently regulated. Far too much of the time it becomes a race to the bottom in terms of language, a lack of basic human decency and respect for others and extreme irresponsibility. All of these behaviours seriously coarsen public discourse and do nothing to inspire trust – especially when so many people are exhibiting the same negative behaviour. The platforms amplify negativity and irresponsible commentaries. They allow bad hats to thrive too often unchecked for too long. It is only in recent days that this is beginning to change for the better.

The social media platforms do not deserve all the blame. We have to accept our fair share too. It sometimes seems many people today have chosen to put their brains on ice. They don't use them. Instead, they are happy to live in echo chambers where their worst instincts and fears are amplified. The modus operandi is: if you are not with us you are against us and, therefore, heretics who must be destroyed. It is all or nothing.

As a result, truth is as often in as short supply as trust is these days. Unsubstantiated rumour and conjecture are taken as the gospel truth because the echo chamber of choice is always right even when it is not. Too few people bother to compare sources of information because alternatives have been ruled out precisely because they are by definition 'wrong' and 'evil'. How can we begin to discuss issues if we cannot agree on the facts of the issues?

The result is too many of us are failing as individual leaders. We are doing ourselves a disservice. We are not role models for others let alone for the next generation. Mouthing off has become a God given, inalienable right. Wrong. Just plain wrong.

If we are to achieve the best outcomes for ourselves, our families, our communities and our countries we need to step back and reflect. We need to answer the question:

'What am I doing to help others?'

In our interconnected and interdependent world we need to care about others as well as ourselves. We need to participate in the life of our communities and do our best to help according to our individual abilities, interests and means.

It is high time social media platforms were regulated so that they are fined if they do not tighten their rules on the tone and behaviour expected of users. Bullies, foul language and other forms of 'nuclear' expression and behaviour should be banned. Positive outcomes should be highlighted.

I am not suggesting for one moment that criticism be outlawed. Far from it, I am arguing that criticism should be constructive not destructive. It is so easy to criticise someone without suggesting what might work better. It is so easy to undermine a fellow human being instead of helping the person see a potentially more positive result which will benefit the community.

I am also not suggesting that investigative journalism be banned. Not at all but like any other user investigative journalists need to take responsibility for their words and the effects their words have on others.

The key points here are accountability and responsibility which are exactly what we rightly expect of corporate and national leaders. Why should we not expect it of platform owners and of all users including ourselves? Our actions and behaviour matter and we need to accept our individual responsibilities for the nature of outcomes.

Does this all sound too much like Utopia to you? Not at all, Utopia is unattainable because we are only human and fallible. We make mistakes and will continue to make them. The important thing is that we learn from our mistakes and take responsibility for our words as well as our deeds. It is the only way forward for us as individual leaders.

Appropriate laws can help protect us and social media platforms which need to be better regulated not only concerning their behaviour and that of their users. They also need to be better regulated to control the use of the data they collect from users.

Before we complain about the leaders we vote into power or report to in companies we need to take stock of our own actions as individual leaders. What are we doing to inspire trust, collaboration and positive outcomes?

The pandemic has forced all of us to adapt and comply with safe (sometimes also called social) distancing rules, mask wearing and to be responsible in the best interests of our communities. It has been remarkable how many people have complied with all the restrictions. They are exhibiting leadership qualities because they accept that their actions and behaviour could adversely affect others.

The minority of people who believe the rules do not apply to them, that they are somehow 'sovereign entities' are being selfish by not caring for the wellbeing of others. They do not accept responsibility for their actions and do not care about other people. They need to be held to account under the law.

At the national level in many countries, too many of us abdicate leadership to others without recognising our responsibilities as citizens to actively participate in the life of the communities in which we live and work. Too many of us reject the notion that we have any power to influence elections. That's for others, many say. What is that old adage? 'You get the leaders you deserve'. A large part of individual leadership is caring about positive outcomes for the community, for the country as well as for the individual.

It is fair to say that trust is in short supply in many companies and countries today because of a failure of leadership in business and in public life. Good ethics and governance have been abandoned by too many people who should know better but who don't or who don't care to know. The temptation of short-term opportunism for selfish personal, corporate or party gain is just too hard to resist. This behaviour is what I call selfish, destructive leadership when what businesses, societies and the world needs now is selfless, constructive, collaborative leadership. We need these qualities because we all face an existential crisis of our own making.

Our planet is at a tipping point. It is overpopulated and we humans have spent the last 80 years over consuming the planet's resources in a wholly unsustainable way. There are always consequences for selfish behaviour. The chief one of which is that, we have acted and continue to act against our own best interests – against our own survival as a species and the very sustainability of our planet. Climate change is an existential risk which we can no longer ignore because our lives and businesses depend on us collaborating to mitigate it. That calls for selfless, constructive, collaborative leadership also known as servant leadership. This term comes from the fact that leadership should never be about the leader or the leader's ego or power for power's sake. Leadership should always be about the leader doing right by and for those people for whom he or she is responsible. In today's world that means a leader's own country and all others prepared to collaborate to find solutions to climate change.

As 2020, has demonstrated the demands on and expectations of leaders is higher than ever. No wonder so many do not live up to either. Like us they are just human. And yet, where leaders strive to always live up to their responsibilities, everyone benefits. You only have to look around the world to see very clear examples of this.

The countries which have managed the pandemic better have leaders who care about their responsibilities for their fellow citizens and residents.

Many are women leaders. I have already mentioned Ms. Jacinda Ardern the Prime Minister of New Zealand. Leaders like her listen to medical and scientific advice. They demonstrate empathy, communicate clearly and take firm action when circumstances dictate. In doing so they are practising true leadership: selfless, constructive, collaborative and decisive leadership in the best interests of all.

When there was an outbreak of COVID-19 because the rules of a quarantine centre were not followed by staff, Ms. Ardern decisively sacked the people responsible and put an army officer in charge. The key responsibility of all leaders is to take appropriate action.

We will never achieve true collaboration among nations until all accept there are, and always will be, different political systems. Provided those different systems do not adversely affect other countries, they have to be accepted as they are and not what others would like them to be.

It has been the commonly held dictum in the developed world that democracy is the best system of governance and the rest of the world should be encouraged to adopt it. The problem is there are multiple types of what passes for democracy and none of them are perfect.

It is not possible to impose a system on any country and expect similar standards and results to those which are meant to apply at home. This always fails wherever democracy is imposed.

A country's political structure and government is a matter for the citizens of that country not for another country to attempt to decide, influence or impose.

That said, democracy has not been doing itself any favours in recent years. The Trump presidency in the United States has been a propaganda coup for other forms of government around the world. It has made democracy much less attractive.

Instead of preaching to others, all democracies need to focus on putting their own houses in order. They need selfless leaders committed to working with their fellow citizens to reform institutions to make them fit for purpose. Selfless leaders who care about the people they represent and are prepared to act in concert with others to strive to reduce and remove economic and racial inequalities. These inequalities are like a cancer which blight communities, increase a sense of exclusion, drive up crime rates and waste human potential.

The same inequalities are the reason why globalisation has been rejected by many who have lost out and feel not only excluded and helpless but that they have nothing left to lose. This creates fertile ground for social unrest which undermines the economy, social harmony and progress. Everyone's quality of life is diminished.

These unintended consequences of globalisation have nothing whatever to do with international trade. They have everything to do with a failure of domestic policy itself a result of uncaring, selfish elites and leaders. That failure must be righted so that societies in the United States and elsewhere can heal and progress.

As a small city state, Singapore has always had to be pragmatic. It accepts the world as it is not what it would like it to be. Singapore seeks to collaborate with all countries prepare to work with it while maintaining its own interests.

The country's political philosophy is utilitarian. Government exists to ensure the best possible outcomes for citizens. That approach took a divided former British colony and turned it into today's global city in one generation. This remarkable achievement was founded on the belief that a multi-racial, multi-religious city-state needed to include everyone in order to progress. That is why Singapore has four official languages: English, Malay, Mandarin and Tamil. English is the language of commerce, government and the courts. It is a common language for all racial groups all of which are equally respected under the law.

Another example of inclusion is Singapore's public housing. Eighty per cent of Singaporeans live in these flats which are well-built and of good quality. Older estates are regularly upgraded in the interests of fairness and to ensure they don't deteriorate.

The government ensures a racial mix in all estates not by apartment block but by each floor of each apartment block. Granted, this most socially intrusive of policies would not be accepted elsewhere but it makes perfect sense for multi-racial, multi-religious Singapore. Indeed, the policy is the result of caring leaders who wanted to build a new, racially harmonious future for the divided city they inherited. And to solve the housing shortage inherited too.

In the corporate world, selfless leadership is demonstrated by leaders who care about their teams just as much as they care about their customers, their communities and their shareholders. The truly successful business leader is one who works with his or her colleagues to co-create and sustain a positive workplace culture. One in which people are respected, valued, developed and included irrespective of race, gender, sexual orientation or any other potential barrier to inclusion. Here in Singapore, inclusion includes bringing local and foreign talent together to learn from one another, to collaborate, to innovate and to achieve desired business outcomes.

The corporate selfless leader spends his or her time removing whatever real or imagined barriers prevent their teams from excelling and giving of their best. This has to be enabled by fair hiring and career paths open to all suitably qualified candidates.

Ultimately, it is the character and behaviour of the leader which makes all the difference. Does he or she walk the talk consistently? Do they say one thing but do another? Do they treat people with respect or bully their subordinates? Are they honest with themselves and others?

In the business world, just as in our individual lives and in political life, decent behaviour and ethics have never been guaranteed. Whether or not good ethics and good governance are consistently practised depends on the leaders and the workplace culture of each organisation. Self-regulation does not work. What is that refrain beloved of auditors? Ten per cent of people are honest and will never defraud, 10% are crooks and the remaining 80% are potential crooks if they think they'll get away with it. This is why there are governance lapses all the time and why every business and society need checks and balances controls and processes. No one is above the law and every leader has the responsibility to lead by example and adhere to corporate governance rules and the law. Every leader has the responsibility of ensuring his or her teams do the same.

One of the silver linings of COVID-19 is that more business leaders than ever before are aware of their own mental health and that of their teams'. This is a really positive development. Is it too much to hope we humans are finally growing up? That we are at long last able and willing to talk about what this and other taboos? Let us all hope so.

The human condition unites us all which brings me neatly back to climate change – the existential risk of our time. Businesses have a big role to play in mitigating climate risk. Not only is it in their own best interests but it is the right thing to do.

Businesses need to win the trust of their teams, customers, investors and communities by working together to innovate to transition the economy from wasteful linear to circular models of production and consumption. Therein lies true sustainability for every business, every human and for our planet. Selfless, constructive, collaborative and decisive leadership always has a purpose. There is no greater personal, commercial or political purpose today than to survive.

All of us humans are leaders. We should not shirk that role. Nor should we shirk the accountability and responsibilities for acting in the common good. We should hold ourselves and those leaders who represent us politically and those to whom we report in companies to the same standards. If enough of us do so, there is hope for our species and for planet earth.

CHAPTER 3

Domino Effect

How Pandemic Chain Reactions Disrupted Companies and Industries

Michael S. Tomczyk

Mack Institute for Innovation Management, The Wharton School, Philadelphia, Pennsylvania

CONTENTS

There is a classic poem that describes how a series of fateful events led to the defeat of King Richard III who fell from his horse and died in battle in 1485:

> For want of a nail the shoe was lost;
> For want of a shoe the horse was lost;

DOI: 10.1201/9781003148715-3

For want of a horse the king was lost;
For want of a king the kingdom was lost.

This classic verse describes how the loss of a nail resulted in a chain reaction where unfortunate events fell like dominoes, leading to the loss of a king and the loss of a kingdom. In 2020, the COVID-19 pandemic created a domino effect that transformed human society and even caused the loss of an actual modern "king."

In a few short months, the crisis changed how we do business, work, shop and conduct our daily lives. Every country and every industry was affected. The dominoes fell slowly in some industries and quickly in others.

The pandemic forced us to adopt social distancing, close physical locations, wear masks, raise plexiglass barriers at cash registers, and restrict crowded venues like bars, workout centres and sporting events. Workers, families and students had to avoid crowded office buildings, retail stores, athletic events, restaurants, theatres and schools. Each falling domino brought a new surprise. Who could have imagined that almost overnight we would have to stop dining at restaurants, stop going to movie theatres and stop attending soccer games?

We were told: You can't go to work. You have to stay at home. You can't dine inside a restaurant, you have to order food or pick it up outside. You can't watch movies in theatres. You can't hold meetings or family gatherings. Your kids can't go to school. There were so many rules! It began to feel like we were trapped in an Orwellian nightmare or a really bad apocalypse movie.

3.1 THE BIOLOGICAL CRISIS CREATED A SURGE IN INNOVATION

If we look at how the pandemic's domino effect played out, we can trace the progression from the onset of the disease to the impact on emerging technologies. As the pandemic slowed down society, it greatly accelerated the use of digital technologies and applications. Many everyday tasks that required physical contact, mingling with crowds, handling money or touching keypads or door handles, were moved online or to smartphone apps. This created a surge in new user applications for computers and smartphones.

Here's an overview of how the "dominoes" led from a biological crisis to a technological renaissance:

Domino 1: the pandemic shut down businesses and forced people to "shelter in place" at their homes,

Domino 2: many companies and stores had no revenues for months and many had to close their businesses or file for bankruptcy protection,

Domino 3: the threat of disaster forced smart, flexible retailers to increase their use of digital/online technologies and applications to stay in business and

Domino 4: innovative cutting-edge technologies such as FinTech, 3D mapping and virtual assistants supported the expansion of telework, virtual meetings, mobile cash, streaming video, virtual stores and online shopping. Many businesses were able to survive the pandemic and keep functioning by switching to digital forms of banking, shopping, working and entertainment.

3.2 A MEDICAL APOCALYPSE

The year 2020 felt like we were trapped in medical apocalypse. We've all seen pandemics in science fiction movies but now a pandemic was really here. COVID-19 was an incurable disease as scary as the Bubonic Plague or the Spanish Flu. The plague killed half the people in Europe in the 1300s and the Spanish Flu infected half a million people and killed 50 million in 1918. From Fall 2019 to January 2021, COVID-19 infected more than 94 million people worldwide and killed more than 2 million. COVID-19 was especially lethal because the virus molecules used protein 'tentacles' to infect cells, requiring new medical approaches.

For want of a vaccine, many people were lost—fortunately, despite the severity of COVID-19, the global medical community was able to diagnose the virus and develop vaccines in record time (less than 6 months). Vaccines developed by Pfizer, Moderna, Johnson & Johnson and other pioneers were the first-ever vaccines to incorporate proteomics, RNA and DNA.

The timing of the COVID-19 outbreak was fortuitous because the medical community had already faced previous pandemics (SARS, Ebola, H1N1) and COVID-19 was actually a type of SARS/coronavirus. Also, giant strides have been made in the past half-decade in the field of genomics, gene therapy, gene editing, proteomics and molecular medicine in general. Unlike previous disease outbreaks, the scientific news and infection updates were available on a daily basis through podcasts, online news reports and YouTube video interviews by leading scientists and practitioners (see Figure 3.1).

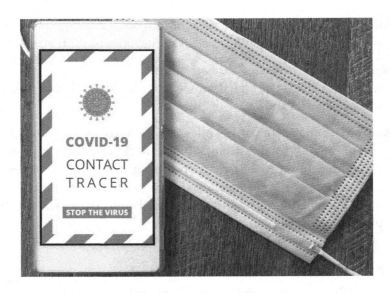

FIGURE 3.1 Digital 'pandemic apps' range from infected victim contact trackers to biosensors for taking temperatures. (Source: istock/Kanyakits)

Several digital applications were also developed specifically to help manage the pandemic such as temperature accessories that attach to smartphones for quick testing to screen people entering buildings. Online applications were also developed that allow the use of smartphones to manage contact tracing, to identify people who came into contact with someone who contracted the COVID-19 virus.

3.3 ECONOMIC DOMINOES THAT CRIPPLED COMMERCE

In addition to the 'medical dominoes' that caused so many infections and fatalities, there were 'economic dominoes' that created real-life nightmares we've only seen in Apocalypse movies. People were forced to stay at home. They couldn't go to work. As restaurants and other businesses were shut down, workers lost their jobs and sources of income. The largest shopping malls in the world were forced to close. Property owners lost rent. Governments responded by providing emergency loans which forced small businesses to go deeper into debt to cover their costs when no revenues were coming in. After several months with no income, many businesses went bankrupt. Worldwide, businesses that were forced to close permanently were not thousands or tens of thousands, but *hundreds of thousands.* By Fall 2020, more than 100,000 business were permanently shut down in the United States alone.[1]

The real challenge to the business community was how to keep selling goods and services without contributing to the spread of the virus. We had to invent new ways of communicating, new ways of selling things, new ways to shop, new ways to safely visit restaurants and banks and athletic events. We had to wear a face mask at the supermarket and disinfect the shopping cart handles before touching them. We had to remember not to touch our face after going to get gas or pressing the touchpad at a cash register. We had to communicate with friends and family by video chat instead of visiting them in person.

Most people couldn't go to work because offices and factories were shut down except for essential businesses like food processing and supermarkets but supermarkets had rules, too (Figure 3.2). There were

FIGURE 3.2   Michael S. Tomczyk checks out the pandemic safety warning at a local supermarket where early morning hours are reserved for high risk shoppers. (Source: Self-photo by M. Tomczyk)

special hours for elderly and at-risk customers, and mandates to keep six feet away from other shoppers. You could get yelled at for standing too close to someone in the checkout line and worse than that, you could contract a fatal virus just from going to the supermarket. The world was suddenly a bizarre and dangerous place. Standing next to someone who sneezed or coughed could be fatal. Some people could be sick and exhibit no symptoms so they seemed safe but were really dangerous to be around.

Things we took for granted like sitting around a table in a small conference room at work was now forbidden. We had to meet online using a video conferencing application like industry leader Zoom, which reported a surge from 10 million daily users in December 2019 to more than 300 million daily users by mid-2020. Even after the pandemic is eradicated, industry experts predict that going forward, less than 25% of all meetings will be held in-person. The convenience and cost of online meetings has created a 'new normal' for how and where meetings are conducted.

By Fall 2020, audiences on TV game shows and talk shows were no longer allowed to sit in crowded bleachers. They were shown as a matrix of people peering at computer screens from home. Commentators and reporters on TV also gave their reports from home instead of in the studio. Some TV shows had to shut down completely.

In our personal lives, many routine activities were suddenly illegal or impossible, like going to a movie theatre or eating at a restaurant or exercising at a gym. These restrictions presented us with some weird choices. Will we lose weight from having fewer restaurant meals, but gain weight from not being able to visit the gym? Can we touch the package delivered by the UPS driver? Is it okay to hand our credit card to a checkout clerk, or should we insist on swiping it ourselves? Should we shop at a store where lots of people are shopping, or buy more items using Amazon or Alibaba? These are choices that suddenly could be life or death decisions.

Fortunately, the challenges of how to work from home, how to communicate with colleagues and customers, how to hold a meeting, how to shop for something without going anywhere or touching anything … these challenges were addressed by technology! Lucky for us, our mobile phones had evolved into smartphones and broadband data had moved to the Cloud so we could video chat with friends and family, shop online and collaborate online. The explosion of digital electronic media, ubiquitous mobile wireless telecom and the Internet of Things had converged at exactly the right time to help us cope with the COVID-19 pandemic.

3.4 DIGITAL TECHNOLOGY TO THE RESCUE!

The falling dominoes – the negative impacts caused by the pandemic – created a lot of problems, that's true. Much of normal life and work stopped. But these problems also accelerated the adoption of digital technologies and applications that allowed businesses to keep operating. Digital apps made it possible to keep watching movies on streaming video, increased the use of online banking and online shopping, exploded the use of home delivery services and much more. Digital technologies turned out to be a lifesaver for companies and industries as well as at-risk individuals who need to avoid physical contact.

In many areas of our lives, the pandemic forced radical changes that will continue to affect our lives for decades to come. Following are a few examples.

3.5 WE ARE NOW TIME-SHIFTERS AND PLACE-SHIFTERS

One of the major changes in behaviour accelerated by the pandemic was the shift from 'on schedule' to 'on demand' services. Instead of being required to access a product or service at a time or place determined by the seller, it's easier than ever for the BUYER to choose the time and place for delivery. Instead of going somewhere to buy something, we buy something from an electronic catalogue and the item is delivered directly to us, often with free shipping. We now order a widget from Amazon or Alibaba and we even choose the date it will arrive at our front door. We decide the place so in a sense we have become *place-shifters*.

We are also becoming time-shifters. We can watch movies and TV shows any time we want, instead of being forced to go to a theatre or to watch our favourite TV show once a week. The pandemic lockdowns caused more people to watch streaming video at home which increased our use of media networks such as Netflix, Amazon Prime, YouTube, Disney, HBO Max and Hulu. In 2021, the first run blockbuster movies were scheduled to be released on streaming video the same day they are released to theatre chains. Many TV series are now released as a package that includes all the episodes so instead of watching shows over a months-long season, we can binge watch the entire series in one weekend! In this sense, the pandemic has given us a power that sounds like something from a science fiction movie: we have become *time-shifters!*

3.6 MAKING FILMS AND TV SERIES IN THE PANDEMIC ERA

In the entertainment industry, movie fans have been transitioning from movie theatres to streaming video for decades. The pandemic closed theatres where side-by-side seating in auditoriums created potential virus traps. When theatre chains were forced to close for months, many producers had to delay filming and change their practices to keep actors and crews safe. Several TV shows that featured audiences in studios went on hiatus. Late night comedians started broadcasting from home. A few shows were forced to be extra creative. In the United States, the pandemic forced the popular series BLACKLIST to *stop filming* in the middle of the last episode of Season 7. To complete key scenes the producers substituted animation for live action which was only needed for a few scenes. This allowed them to complete their final episode.

The impact of the pandemic on the movie and theatre industry has been profound. During 2020, the biggest movie chains were closed down completely. Cineworld closed its theatres early in 2020 and again in October 2020. Regal Cinema furloughed 24,000 of their 25,000 workers in the US and as many as 40,000 or more worldwide. AMC laid off 26,000 of their 27,000 employees. Historically, the film industry has relied on revenues from theatre ticket sales to fund high budget blockbuster films that can cost tens or hundreds of millions of dollars to produce.

When theatre chains were forced to close in 2020, many studios cut back or delayed filming of blockbuster films which relied on massive theatre sales to cover their costs. Many films that were scheduled for release in 2020 were pushed back to 2021 as a result of theatre closings as well as production delays imposed by pandemic safety rules. One example is the James Bond film, *No Time to Die*, which was pushed back from Fall 2020 to April 2021.

In Fall 2020 Warner Brothers announced that all 17 of their first run films planned for 2021 would be released simultaneously in theatres AND streaming video (HBO MAX). These films include *Matrix 4* and *The Suicide Squad*. Each of these films will be available on HBO MAX for 31 days from the theatrical release date. The cost of most first run films viewed on home cable services cost the same or more than theatre tickets. On a positive note, the audiences for first run films offered on streaming video will be exponentially larger than in-theatre audiences, so film studios may stabilise, financially, although all movie theatre chains have already suffered huge financial losses.

3.7 MOBILE MONEY: THE CASHLESS SOCIETY

The evolution of a cashless society is being accelerated by the need to avoid handling physical money and the risks associated with visiting bank lobbies, teller windows and ATMs where the virus can be transmitted by physical contact.

In 2020, there was a surge in new apps and services called 'FinTech' – financial technologies – that changed how money is defined and moved. Currently, traditional banks charge high fees and impose long processing delays to move money across borders from one country to another. For example, when $1000 is sent to a family overseas, only $800 may reach the recipients after high fees and currency charges are applied. The United Nations has recognised this as a problem and has set a 'sustainable development goal' to reduce migrant remittance fees from 7% today to 3% by 2030. The demand for more mobile money applications has created a surge in FinTech startups.

A US-based venture called FamaCash™ (which I have been privileged to advise) is a mobile money network that allows people in virtually any country to transfer money across borders faster and at lower fees. FAMA uses advanced technologies to reduce costs and speed up transactions. FAMA's mobile money platforms use blockchain for recordkeeping and uses a US dollar secured cryptocurrency called stablecoin which is essentially 'digital money'. FAMA also uses artificial intelligence and cloud computing. Mobile money can be transferred across borders in more than 100 countries, and seamlessly converted among more than 50 different currencies.

Advanced technologies are redefining the concept of money from paper bills and metal coins to digital currencies. Mobile money can be stored and transferred across borders using a smartphone, which means anyone with a mobile phone can send or receive money, internationally. The use of smartphones to transfer money is projected to become a multi-trillion dollar industry. The untapped market is enormous. For example, more than 2 billion people in the world do not currently have access to a bank but own a mobile phone. These people need financial services and the convergence of smartphones and mobile money apps are just beginning to provide solutions.

A family living in a remote region in Bangladesh for example, may not have access to a bank or banking services, but most people in Bangladesh own a mobile phone. They can use a FAMA mobile money platform such as Remit™ or Sonali Pay™ to transfer money even if they don't have a bank account.

There are many new applications and services that are being created by the evolution of mobile money. For example, peer to peer payments can now be made using a smartphone. When a group of friends want to pool their money to buy a gift or pay for a group meal, they can use a 'peer to peer' service to move money instantly. One example is PayPal which created an application called Venmo which enables quick real time transfers of money between friends and colleagues, using a mobile phone.

An extension of this application is Peer to Peer Lending, which involves enabling one person to make a private loan to another person, even if the person receiving the money lives in another country. Peer to peer lending becomes more feasible in locations where interest rates are high or where private loans and mortgages are being extended by family members or business colleagues to help out during a financial crisis.

3.8 ELECTRONIC SHOPPING SURGE

Thanks to online shopping, instead of traveling to a department store, sporting goods store or home improvement centre, we can shop online. We can sit in an easy chair in our living room and buy things – avoiding in-store crowds that could give us a fatal virus. Ecommerce has been gaining ground since the 1990s but the pandemic shut down most department stores which meant the only way to sell merchandise was online.

Retail chains that did not cultivate online shopping faltered or went bankrupt. In the United States, the list of companies that filed for bankruptcy protection during the pandemic included such iconic retailers as Brooks Brothers, Chuck E. Cheese, GNC, J.C. Penney, J.Crew, Lord & Taylor, Neiman Marcus, Men's Wearhouse and Pier 1. Lord & Taylor was founded in 1826 and Neiman Marcus was founded in 1907 so these are true legacy brands and their filing for bankruptcy reveals just how powerful the impact of the pandemic has been on even the most venerable brands.

3.9 3D SHOPPING ALLOWS SHOPPERS TO NAVIGATE REAL STORES

Another innovative development that gained traction during the pandemic was virtual store shopping. 3D shopping allows shoppers to navigate through a real store location like we navigate through a video game.

Alibaba has been a pioneer in the testing of virtual reality shopping devices and online store apps. In 2016, Alibaba's Taobao app allowed

FIGURE 3.3 Virtual reality allows online shoppers in China to visit and navigate Macys' New York store and other store locations using Alibaba's Taobao app. (Source: Alibaba)

Chinese shoppers to explore Macy's retail store in New York (Figure 3.3). The accompanying image shows an online 3D map of a Macy's store in New York that was made available to Alibaba shoppers as part of a market test. The arrows on the floor are used to navigate through the store and the circles can be clicked on to expand a store shelf or rack to show a selection of shirts, pants, appliances and other items which can be purchased online for delivery.

In May 2020, Alibaba pilot tested Luxury Soho, a platform named after New York's Soho district. These innovations allow shoppers to visit actual store locations without leaving their living rooms and are one of the many ways retailers are working to overcome store closings imposed by the pandemic. If we can't go to the retail stores, some retail stores are bringing the stores to us! This is a unique form of "Placeshifting" adapted from the video game industry where virtual reality is already well established, with over 250 million users worldwide.

3.10 THE RISE OF TELEWORK

When I was working as Managing Director of the Mack Institute for Innovation Management at the Wharton School (I retired in 2013) I remember the day the Wharton Accounting Manager came into my office and proudly announced, 'I'm working at home one day a week from now on'. That was my first encounter with what we now call 'telework'. In academia, faculty and program managers were used to working at home, but for people working in accounting and other areas, telework was something new and special. It meant they could do office tasks at home. When the pandemic hit, telework became a necessity because many office buildings and plants were closed. Also, kids were home from school and many parents had to stay home.

Fortunately, we had already created a 'connected society' where everyone can communicate with everyone else thanks to digital data and wireless smartphones. Applications and computer files are stored on Cloud servers which can be accessed from smartphones, laptops, iPods and other smart devices. We can text each other, exchange photos and videos, and chat face to face in real time, bypassing traditional telephone lines. We use our smartphones to exchange messages, move money, shop online and get movies, meals and taxi services on demand. Physical documents can be digitised and shared and edited online using Google Docs, Dropbox and other applications. We hold meetings online using Zoom instead of meeting in small conference rooms where the potentially fatal virus can be spread. The pandemic has accelerated telework, and on balance, we are becoming more efficient and productive than any time in history.

3.11 POLITICAL DOMINOES: FOR WANT OF A BALLOT

In the United States, a historic chain of events occurred when President Donald Trump lost the November 2020 election to Joe Biden. It can be argued that the pandemic crisis (and to some extent, digital technology) played a role in this loss – for want of a cure, a president was lost?

When the coronavirus epidemic exploded in America, the medical community scrambled to discover the best way to prevent the spread of the virus. Communication was virtually instantaneous thanks to online news that virtually everyone could access via their smartphones. However, the Centers for Disease Control issued conflicting advice concerning wearing masks, the danger of contacting the virus from touching surfaces and door handles, and which existing drugs might be effective. In many states, governors declared statewide lockdowns. Some states made it mandatory to wear masks, others did not. As a result, the virus spread rapidly, pressuring medical facilities. Many Americans viewed the mandates to wear masks as an infringement on their freedoms. President Trump was seldom shown wearing a mask.

On September 26, 2020, the president presided over an event in the White House Rose Garden to announce the appointment of Supreme Court Justice Amy Coney Barrett. The 150 people attending the indoor and outdoor event were told they did not need to wear masks and side-by-side chair seating did not allow social distancing. Subsequent to this event, dozens of prominent officials contracted the COVID-19 virus including President Trump and his wife, two US senators, numerous White House aides and advisors, the presidential press secretary, Secret Service agents, academic leaders and several reporters. Also, during the presidential campaign, hundreds of people who attended Trump rallies were shown not wearing masks.

It can be argued that the somewhat cavalier attitude towards wearing masks and social distancing demonstrated by the president and his staff influenced the November 2020 presidential election. The president's handling of the crisis became a 'domino' when voters saw that President Trump did not seem to be taking precautions himself, when he was supposed to be the nation's role model for virus prevention. Also, many voters resisted state mandates to wear masks and saw requiring masks as an infringement on their freedom. There were daily videos and news reports showing confrontations with people refusing to wear masks in public. While the president was directly responsible for expediting vaccine development, which was an historic medical achievement, his handling of virus prevention in the White House and at campaign events conveyed mixed signals to the public.

3.12 FOR WANT OF A BALLOT…

On November 3, 2020, Joe Biden defeated Donald Trump and was elected President of the United States and the President's handling of the COVID-19 virus outbreak was certainly a factor.

One critical political "domino" in the election was the adoption by many state leaders of new election rules that accepted mail-in ballots, allowed counting of ballots as long as a week or 10 days after the election, and some states enabled ballot harvesting. These rule changes were adopted to give voters an option to vote without physically attending a crowded polling site where they might contract the virus. Mail-in balloting and automated processing was enabled by several digital technologies that processed votes, signatures, and electronically verified names and dates.

Republican voters claimed that the new rules were illegally imposed, favoured Democrats, and facilitated ballot fraud. After the election,

President Trump and his allies filed more than 60 state and federal lawsuits, virtually all of which failed.

The widespread claims that the election was 'stolen' by Democrats culminated in a massive rally attended by tens of thousands of Trump supporters who convened on January 6 near the Capitol building in Washington D.C. This event has been called a rally, a protest, a riot and an insurrection. Following speeches by Trump and key supporters, hundreds of people marched to the Capitol building, broke through security lines and invaded the building, where they vandalised offices and voiced threats against legislators who had to take refuge from the mob. The capitol invasion led to the second impeachment of the President by the Democrat-controlled House of Representatives.

Ironically, throughout 2020, Democrat mobs of rioters had invaded and vandalised federal buildings in several US cities including Portland, Seattle, Minneapolis and many others. The January 6 invasion of the Capitol complex totally reversed the narrative and enabled Democrats to label Republicans as the violent mob who invaded federal property. The attack on the Capitol also gave social media giants like Facebook, Twitter, Amazon and YouTube a reason to censor and ban conservative posts, pages, websites and even networks.

Looking back on the 2020 election, the 'dominoes' included sloppy handling of pandemic precautions by national political leaders, new ballot handling rules that favoured Democrats in key states, the loss of the presidential election in November, court cases that failed to prove election fraud, angry mobs protesting what they saw as a 'stolen election' which led to the invasion and trashing of the Capitol, which then provided Democrats with a reason to impeach President Trump, and also gave social media an excuse to censor and ban conservatives.

Whether you agree or disagree with the political outcomes, it is clear that the pandemic did in fact contribute to the 'loss of a king' or rather, the loss of a president.

3.13 WHAT COMES NEXT?

As outlined here, the COVID-19 pandemic has created a wide variety of chain reactions and 'dominoes' that have threatened and transformed our society and everything we do. Public events, news stories, technology successes and failures, medical innovations, political responses to the pandemic, legal decisions and more will determine which transformations are permanent, which are temporary and which will have lasting effects.

I predict that as the pandemic is eliminated thanks to vaccines and/or cures, society will return to a new normal where the use of technological innovation will continue to give us increased productivity and make our daily lives safer and more hopeful.

We will continue to see new rules, laws and protocols put in place to give decision makers more control to deal with crises like a pandemic. The yin and yang tug of war between too much government control and not enough will continue, and this has always been a historic conflict. What we are learning from COVID-19 is that it is essential to keep innovating in all areas, from digital technology to molecular medicine, because there are forces at work in the world that have the ability to destroy our societies, cultures and human species.

My strongest takeaway from the COVID-19 pandemic is that this global crisis has given us a worldwide wakeup call which hopefully means we'll be better prepared to adapt to any future crisis – whether it's a pandemic, asteroid collision, super volcano, war or other threat – that creates new domino effects.

NOTES

1. Nearly 100,000 establishments that temporarily shut down due to the pandemic are now out of business; by Anne Sraders and Lance Lambert, FORTUNE Magazine, 28 September, 2020

CHAPTER 4

Strategic Reframing for Retailing in a Post-COVID World

A Scenario Planning Approach

Malobi Mukherjee

James Cook University Singapore, Singapore

CONTENTS

DOI: 10.1201/9781003148715-4

4.1 INTRODUCTION

The retail industry worldwide has succumbed to the unprecedented challenges posed by COVID-19. Robinsons in Singapore and Debenhams in the United Kingdom are just two examples of iconic stores that have joined the long list of well-known retailers called into administration due the sever impact of the pandemic. The socioeconomic pressures of the pandemic were the final nail in the coffin for a number of these retailers with pre-existing strategic and operational issues which had already made them irrelevant and uninteresting to customers (Treadgold and Reynolds, 2020; Mukherjee 2020). So what lessons can future leaders learn from this turn of events? In this chapter, the author proposes the foremost lesson for retail leaders will be to revisit existing approaches on risk assessment and consider alternative approaches such as scenario planning to develop resiliency from future shocks. An approach worth considering is that of scenario planning which offers a framework to engage with external disruptions, raise plausible, counterintuitive questions about the future from the vantage point of the present and challenge a retail executive's fundamental assumption about their businesses. Using the inductive scenario planning methodology, the author proposes three distinctive scenarios which surface fundamental changes retailers would need to undertake to future proof their businesses from macro environmental shocks in the future.

The layout of the chapter is as follows. First, the author provides an overview and justification for strategic reframing with scenario planning. This is followed by a discussion on the various scenario planning methodologies with a particular focus on the inductive scenario planning process. The next section presents a number of post-COVID mega trends and the combined influence of these trends have been used to develop future scenario stories. In the penultimate section, the scenarios are used to raise counterintuitive "what if" questions to deliberate the transformative roles key retail stakeholders would have to assume if any of the scenarios were to present themselves. The resultant implications for strategic reframing are discussed in detail before concluding the chapter.

4.2 SCENARIO PLANNING AND REFRAMING

Scenario research is a rigorous and practical interactive inquiry process (Ramírez et al., 2015) and used to enable stakeholders to frame and reframe their current situation. This reframing is articulated through a small, bespoke set of manufactured plausible future contexts (the scenarios themselves) which are produced in scenario research workshops facilitated by scenario specialists. The future scenarios are used as safe conceptual spaces in which to consider current decisions, (c.f. Wack, 1985; van der Heijden, 2005) address current concerns, reframe and re-perceive possible courses of action (Mukherjee, Ramirez and Cuthbertson, 2020) and compensate for the usual overconfidence and tunnel vision in decision making (Schoemaker, 1995).

The author adopts the scenario planning approach to develop three plausible, relevant and challenging scenarios for the retail sector in a post-COVID world to raise some counterintuitive questions about the future of the retail sector, from the vantage point of the present day. One of the key lessons of the COVID-19 pandemic was that retail leaders would need to revisit existing approaches on risk assessment and consider alternative approaches to develop resiliency from future shocks. The purpose of developing these scenarios is to challenge a retail executive's fundamental assumptions about the business post COVID-19, consider multiple ways in which the future could unfold and thereby have robust and agile strategies in place to cope with the same.

4.3 INDUCTIVE METHOD OF SCENARIO BUILDING

In this chapter, the author has used the inductive approach to develop scenarios using a combination of desk research, document and online archival analysis (Van der Heijden, 2005). Examples of scenario studies based on desk research include Bobbitt's "The Shield of Achilles" (2002), Schwartz et al's "The Long Boom" (1999), and McRae's global scenario for 2020 (1995). The author adopted a similar approach by using secondary research to identify novel, uncertain and disruptive mega trends in the macro environment. These mega trends have been combined to form a set of drivers of change and their combined impact on the business landscape for the key players in the retail sector have been discussed. The following sections of the chapter outline the process adopted by the author.

4.4 MEGA TRENDS & CRITICAL UNCERTAINTIES IN A POST-COVID WORLD

4.4.1 Global Economy and Geopolitics

The global impact of coronavirus has raised some fundamental questions about a seismic shift in balance of political and economic power from the West to the East. COVID-19 has indeed been that watershed moment and it is almost certain that there will be no return to the pre-COVID-19 era. The question that remains to be answered in the future is the behaviour of nation states, governments and leaders and whether they will work together more closely or whether the trauma of COVID-19 divide them further (*The Guardian*, 2020). One view is that the pandemic will strengthen the state and reinforce nationalism as governments will prioritise national needs over a global outlook and open trade borders cautiously and adopt emergency measures to manage the crisis (KPMG, 2020). Walt (2020) reiterates that the world will be less open, less prosperous and less free with the possibility of state-owned public enterprises making a comeback while governments try to overcome the economic and financial challenges posed by the pandemic. Chatham House in the United Kingdom also raises the prospect that political leaders may retreat into overt geopolitical competition (Niblett, 2020) while the Eurasian Group suggests new and re-merging global divisions between power blocs will give rise to separate supply chain and military powers. The International Crisis Group (2020) has raised a pertinent point about the political implications of COVID-19, suggesting that unscrupulous leaders may even exploit the pandemic to further their own agendas thereby exacerbating domestic and political crises by escalating conflicts with rival states. Views coming from geopolitical experts therefore suggest that global geopolitics in a post global world will contribute to substantial uncertainty in the near future.

4.4.2 Climate Change

The special report by the Intergovernmental Panel on Climate Change (OECD, 2020) made it clear that the previously stated 1.5°C of warming continues to carry significant risks for the global community. The World Economic Forum or WEF (2020) has also stated that despite efforts to achieve net-zero by 2050, global emissions are still rising. Temperatures hitting record levels will result in increased occurrences of droughts and wildfires (ESPAS, 2015) devastating cyclones will hit countries across the world and the OECD (2020) report suggests that yet another global catastrophe awaits us if we let climate change rise unchecked. The global community requires a 'multilateral, strong, resilient, green and inclusive recovery from COVID-19' (Sanchez, 2020) but the road ahead is challenging because the overall funding for a green recovery is still in favour of less sustainable options. More importantly, the challenges facing the planet transcend national boundaries and overcoming them would require consensus and coordination across policy makers, businesses, organisations and communities worldwide to speed up the transition to a net-zero future which in itself is a complex issue.

4.4.3 Technology and Cybersecurity

The World Economic Forum (2020) report on cybersecurity states that by the year 2025 the next generation technology on which the world will increasingly rely, will have the potential to overwhelm and wreak havoc in an increasingly tech-enabled world. In the past decade, cybersecurity has emerged as one of the most important systemic issues for the global economy. The Federal Bureau of Investigation has reported a 300% increase in cybercrimes since the beginning of the COVID pandemic. The World Economic Forum projects a $432 billion global spending on cybersecurity by the year 2030 but it also raises a number of pertinent points about the implementation and efficacy of these measures – would companies be able to afford the increasing cybersecurity costs in a global recession? Risks associated with cyber-threats are often opaque and how would resultant divergent regulatory and organisational approaches enable cross border data flow and e-commerce? The uncertainty in tackling cybercrimes would emerge from the complexities of managing these crimes at a global level and at the appropriate levels of calibration.

4.4.4 Consumer Trends

The consumer trends forecast for a post-COVID world depict the most sobering effect of the pandemic yet – the BBC towards end 2020 estimated that 265 million people around the world would suffer acute hunger because of the pandemic and job losses at the end of the same year were reported to be touching 165 million. A recent report from the United Nations has also indicated that this crisis could push almost 490 million people in 70 countries into poverty (Economist, 2020). A more dismal future is portrayed for the youth who would not only have suffered in the short term following the pandemic but the effect of the pandemic will affect their consumer economic health and spending ability for the foreseeable future.

On a more positive note, new trends in consumer habits reported in Accenture's global COVID-19 Consumer research (2020) suggest that 94% of consumers have made at least one healthy change since the pandemic began – for example, to shop more health-consciously, exercise more or looked after their mental well-being. Some 70% are likely to continue with more than one of these changes. The World Economic Forum (2020) report on the future of consumption also states that consumers are increasingly expecting to develop relationships with, and share responsibility for the businesses they frequent and the people who work there. As per this report, some 56% of consumers have increased their patronage of local shops – and 80% of that group plan to continue the practice. Some 65% of consumers expect businesses and governments to "build back better" – sustainably and responsibly and they are taking personal action to give traction to this movement. Meanwhile, 63% of consumers reported that they are personally limiting food waste by planning their shopping decisions more carefully and using up what they purchase.

The consumer trends therefore portray the future consumer as economically deprived and locally/regionally and sustainably focused. This could present an opportunity or a threat depending on strategic stance of relevant organisations.

4.4.5 Societal Trends

A number of perspectives have emerged on the evolution of society in a post-COVID world. The commonly held view is that the COVID-related adaptations will accelerate already existing trends, like the development of a cashless society, the increase in remote work, the decline of brick-and-mortar retail and some of these will become a more permanent part of the post-pandemic's "new normal". An even more cynical view comes from psychology experts who opine that the traumatic stress caused by the pandemic will affect important milestones including marriages, jobs, dating, entering and leaving schools leading to introspection and reprioritisation of life goals including focusing more on family (Powell, 2020). As physical distancing accelerates in the near future, advances in videoconferencing and other remote technologies will allow many to continue to produce – and collect a paycheck – working from home. The autonomy and choice which remote would bring would also bring a degree of satisfaction in the lives of people (BBC, 2020). An alternative perspective on societal trends highlights the innately social nature of humans and Emanuel (BBC 2020) points out that the debilitating isolation imposed by COVID-19 will trigger an euphoric shift to hyper-socialisation when lockdown restrictions are eased. The testament to this sociality of humans was evident towards the end of 2020, when a number of countries eased lockdown restrictions temporarily, resulting in people rushing to share meals, share a drink in a pub, arranging parties and pursuing outdoor activities. If this trend were to follow in the future, contrary to expectations, a post-COVID world would witness a surge in travels and thriving businesses for all social venues. It may therefore be prudent to assume that cautious optimism would prevail in a post-COVID world where the conundrum between staying safe and returning to a (new) normal way of life would create dilemmas at the societal level.

4.5 FUTURE SCENARIOS IN A POST-COVID WORLD TO THE YEAR 2035

4.5.1 "It's My Life" Scenario

4.5.1.1 The World in This Scenario

In this world, individuals take back control of their lives and choose their respective levels of participation in a digitally connected world. Complete proliferation of technology in everyday lives to maintain contactless interactions following the 2020 pandemic; frequent cyber breaches and cyber security threats to essential and non-essential services and a Cyber 9/11 event during the mid-2000s led to this world coming about in 2030. At the same time, social distancing and long periods of lockdown resulted in individuals sacrificing their health and wellbeing in the digitally connected world of the 2000s and they starting noticing the effect of continuous connection and inactivity on their health and mental wellbeing. The growing digital scepticism of the 2000s rapidly translated into complete alienation from digital solutions provided by the big tech powerhouses of the 2000s.

In the year 2030, Individuals have moved away from transacting with larger organisations due to loss of trust in digital platforms. Communities have started utilising their local resources for trade and manufacturing and this has resulted in a resurgence of local manufacturing and employment. Local, regional and community based start-ups have become the new norm as people want to trade within their community. There is emphasis on protecting and preserving the local and regional environment as citizens settle into regional community life – local resources are utilised prudently, waste is minimal and recycled. Technology is curated for the local community by local businesses. Within each community, the technology platforms are sophisticated but have inbuilt flexibilities. In this scenario infrastructure is fragmented and each infrastructural hub is developed by local and regional players to cater to the unique requirements of the communities living in those regions. The level of sophistication of these infrastructures is dependent upon the financial prowess of the regional local authorities and a network of key stakeholders who form the unique ecosystem within which infrastructural decisions and implementations take place.

4.5.1.2 The Retail Industry in "It's My Life"

The retail industry is focused on value creation derived from flexible solutions that can be curated to meet individual tastes and preferences of citizens who live, work and transact within their trusted communities. Retailing is dominated by small and medium enterprises within the local communities and catering to all the fundamental needs of consumers within those communities. The retail winners in this scenario will be the small businesses catering to the needs of the local communities. Local trusted brands are the first and foremost choice for individuals. The losers in this retail scenario will be the big tech giants, large-scale manufacturers of goods and services, and providers of standardised solutions' a lot of whom struggle to remain relevant in a local community based scenario. Community-based retailing will be dominated by local market circles representing a revolving market square, where the central market stall will be stocking everyday items. Both the central everyday stall and the various occasional stalls operate under their own separate brands, and are usually family run enterprises.

The retail experience is not just limited to community shopping but the entertainment value of the erstwhile urban malls and departmental stores which survived the pandemic of 2020 are an integral part of the experience. Malls and departmental stores have a blurred mix of indoor-outdoor retail space. The service system is geared to a large extent towards providing entertainment for consumers – street performances, entertainment shows, on site craftsmanship and a plethora of services to enhance a greater feeling of community public space are key service elements in this scenario as well. The logistics operators are a network of community representatives facilitating and managing the transportation of goods and service between communities and malls/departmental stores. Local logistics entrepreneurs would form a trust-based supply chain network with stakeholders providing the last mile connectivity playing a crucial role in facilitating movement of goods.

4.5.2 "Heal the World" Scenario
4.5.2.1 The World in This Scenario
This is a world where nation states are experiencing resource constraints and challenges that they need to address pragmatically because of economic, population and demographic megatrends. The combined pressures of population growth, economic downturn post-COVID-19 and climate change will place increased stress on essential natural resources (including water, food, arable land and energy).

The events of the 2000s have resulted in governments making hard choices to protect national economic growth and meet the expectations of the individual national electorates by the year 2035. Government investments are prioritised on the most urgent requirements. In the erstwhile Western developed nations, older populations are requiring more care, new immigrants require more welfare and training and overall government policies including funding priorities are designed to assist in the societal integration of immigrants. New migrants from nations without a history of a welfare state are gaining access to training and family subsidies in addition to health care. Famine, disease, high infant mortality and education continue to challenge an otherwise high-potential population group.

A satisfice approach is taken in this world and the emphasis is on careful, considered long-term choices compared to future investments in urban environments. Existing infrastructures are utilised for as long as possible. Communities value and support governments that make long term investment decisions and are prepared make allowances for some infrastructure limitations that do not impact primary concerns relating to food, health and safety. The needs of the environment and climate change considerations are part of the decision-making trade off. Investment decisions are pragmatic and prioritised for cheap and easy to use smart technologies to address the pressing societal needs. Health and border protection concerns are likely to have higher priority as areas for technology investment because of the lessons learnt from the COVID-19 pandemic of 2020.

4.5.2.2 Implications for the Retail Industry in "Heal the World"

This is a resource-constrained world where retailing is synonymous with conservation of products, usage of second hand goods and sharing/rental services. In this world, retailers find innovative ways to prolong the life of products and services to create value for their customers. The retailers with businesses modelled around sustainability principles valuing the natural environment, conservation and preservation of natural resources will emerge as the powerhouses in this world. Having invested in sustainable and renewable operations in the 2000s, these retailers will have recovered the costs from investing in these green operations and be in a stronger position to realise operational economies of scale. These retailers will have the competitive advantage in selling "value for money" sustainable products and services. The losers in this scenario will be the erstwhile mass retailers who focused on selling cheap products and services globally, with no regard for responsible use of natural resources. These retailers would succumb in a natural resource constrained world. In this scenario the predominant stores would be thrift shops, second hand shops, shared and rental products and services as well as mass scale sustainable, organic and green products. After sales service and repair will thrive in this retail scenario and complement mainstream retailing. Consistency in the consumer experience would stem from familiarity and trust in the brands embodying the sustainability principles in a resource-constrained world; consumers would seek a familiar experience in the purchase and consumption of these brands on every occasion. Resource intensive, luxurious and hi-tech experiential retail malls will be in the past in this scenario. The shine and glitz of unsustainable energy guzzling malls and departmental stores of the 2000s will be reduced to unoccupied derelict buildings at worst, and dark warehouses at best, in a resource-constrained world. Green malls and departmental stores built in the 2000s will be the survivors in this scenario since their infrastructure will enable renewable energy, recycled water and natural lighting. The retail tenants in these green buildings will be an eclectic mix of sustainability-focused stores and brands. Delivery logistics and reverse logistics will be equally important in this world. All retailers will be required to avail reverse logistics facilities as per regulatory requirements. Customer will also have to partake in recycling due to regulatory requirements.

4.5.3 "Robot Rocks" Scenario

4.5.3.1 The World in This Scenario

This is a world where the global big tech businesses (at the size and scale of businesses like Facebook, Uber and Google which existed in the 1990s and 2000s) have grown exponentially and become extremely powerful. The environmental initiative of the global tech giants from the 2000s has slowed down the effects of global warming.

By 2035, AI and IoT technologies are being produced at large scale by the few powerful global tech giants making them available for mass market consumption. These technologies have been leveraged to create an ecosystem of interconnected high quality technology with the physical world that supports all aspects of urban life. Interconnected technology has permeated all aspects of life: mobility, health, education, work and communication with the sole focus of optimisation in every aspect of life. The balance of power is in favour of those countries where the tech companies have their head offices. Governments no longer hold the power and the alliances they achieve with the big tech businesses determine their economic strength. In this scenario, societal trust is placed on the 'city system' for the efficiency and effectiveness it has brought into the lives of citizens. Tech enabled alternative green energy resources have given rise to a cleaner environment where individual wellness and longevity has improved. Since household chores are performed by technology enabled robots and bots, citizens have more time for leisure activities. Technology is built into every aspect of the urban infrastructure to support the ongoing proliferation of tech-enabled every day SMART urban life. Cities have become the hub of work, leisure and living. Technology enabled education has also given rise to the phenomenon of flexi-time schools from flexi-locations. Data is king in this world as the proliferation of technology in everyday lives means individuals have surrendered their privacy to the tech giants managing their smart everyday life. AI enabled technology is controlling algorithms making human intervention and human cyber hacking next to impossible. However, AI has become extremely powerful and the risks of rogue

AI creating havoc and destruction has also risen to unprecedented levels. There is a realisation for the need of increased AI policing, the development of which is at a nascent stage.

4.5.3.2 Implications for the Retail Industry in "Robot Rocks"

Convenience and ease of shopping is of utmost value in this scenario. Efficiency is a key success factor for the retail sector. The variety-seeking, frivolous and dynamic consumption habits of the mid 2000s is replaced by consistent, predictable mass consumption of goods and services. Consumers derive value out of efficient and effective services where choice is limited and standardised. All retail and service providers who have mastered the usage of tech enabled smart solutions are the winners in this scenario. The erstwhile small businesses focusing on production of niche, exclusive artisan type goods and service have lost relevance in this world.

In Robot Rocks, consumer retail experiences occur through customised apps on mobile devices and the interactive use of technology in stores and in malls. The physical stores and malls function more as showrooms and/or warehouses or hi-tech platforms where robotic precision provides a seamless service in unmanned stores to interested customers. Scope for creativity and customised services is limited. The service system is geared towards mass customisation with accompanying click and collect delivery services controlled by super intelligent computers (technological singularity) and robotics technology with utmost precision and consistency. Retailing entails a complete integration of physical and digital channels with inbuilt modular designs allowing customers to access anything from anywhere. The digital interface makes use of transformative technologies such as virtual cloud storage, big data analytics, virtual reality and artificial intelligence to facilitate the ordering of products and services. Big tech companies dominate the global logistical industry. The big tech players invest in collaborative global logistical networks and cloud-based warehousing facilities enabling the purchase and delivery of products anywhere in the world. Logistics are be dominated by drones and driverless vehicles are fitted with 3D printers for mobile production.

4.6 STRATEGIC REFRAMING FOR THE RETAIL SECTOR

The three scenarios proposed in the previous section create the transitional future space within which the proposed courses of action for the retail sector can deliberated, from the vantage point of the present day (Mukherjee et al., 2020). Each of the scenarios pose counterintuitive "what-if" questions for the key players in the retail sector. The alternative or reframed retail perspectives inherent to each scenario challenge the present day business assumptions of retail sector stakeholders and enable re-perception of business roles for improved resiliency if any of the three future scenarios were to unfold, separately or simultaneously.

4.6.1 Strategic Reframing for Retail Real Estate (Mall and Department Store Owners)

The three scenarios in this chapter raise the question "what would be the purpose of malls and departmental stores if future retail experiences were focused on the 'self' or 'community' in 'It's my Life' or 'Heal the World' scenarios or value-seeking, simplified and consistent experiences similar to a 'Robot Rocks' scenario?" Retail real estate developers would need to operate as 'providers of experience' compared to the present day operating assumption as being 'providers of square footage rental space' for stores. Strategically, retail real estate developers would have to reframe their existing tenant-landlord relationship with retailers and work collaboratively with retail tenants as long- term partners with whom they would design and deliver customer experiences. Operationally, the collaborative partners would seek to enhance and sustain relevance of the malls and departmental stores by providing meaningful variety in customer experiences. This would imply alterations and frequent assessment of mall/departmental store retail mix and floor layout with the inherent ability to alter the retail mix based on changes in the overall retail landscape. In the present day, malls and departmental stores operate by segregating food/beverage, apparels and entertainment across different floors. Agile operations in the future could entail further variety in the retail mix to include into community based retail stores, second hand and rental stores, tech-enabled showrooms and basement level dark warehouses for e-commerce. Doing so would prepare mall owners and departmental stores to provide novel, diverse, contemporary and relevant experiences while simultaneously enhancing their resiliency from the proposed future scenarios.

4.6.2 Strategic Reframing for Retail Brands (Apparel, Shoes and Grocery)

Sustainable (green and organic) brands have thus far been positioned as augmented product features targeting medium to higher income consumer groups. The scenarios proposed in this paper suggest that sustainable, green and organic features would be core product features in "It's my Life" and "Heal the World" scenarios, i.e. all mainstream, standardised brands irrespective of their source would adhere to sustainability principles due to regulatory requirements and customer demands. In a Robot Rocks scenario, value for money sustainable products would be created using sophisticated green manufacturing technologies founded on circular economy principles. These scenarios therefore raise the question "*What* role will retail brands assume in the future *if* sustainable products and services become the norm and not the exception?" The most successful retail brands (highest brand equity) would transform their position from creators of sustainable products and services to global educators and innovators of sustainability. As innovators they would be the owners of green manufacturing patents – their mass market brands would manufacture at scale, the niche players would justify uniqueness by controlling the sustainable raw material supplies with their cutting-edge technological prowess. Customer service would focus on educating customers about the principles of conservation and sustainable consumption and loyalty rewards would be based on frequent recycling and reuse of products. The small to medium size retailer brands would lobby for governments to provide green manufacturing sites for use on rental and shared basis. Cooperative style shared green manufacturing sites would use local communities to manage and run operations to support community-based and regional retail brand owners.

4.6.3 Strategic Reframing for Supply Chain Businesses

The three future scenarios create a retail context where retailers irrespective of size would need to operate across modulated platforms to connect the last mile to the urban areas to facilitate local and regional trade in "It's my Life scenario" and enable reverse logistics in 'Heal the World'. The consistent and standardised global transactions warranted by a Robot Rocks world would require a combination of near and far shoring enabled by seamless collaboration between supply chain stakeholders. Retail supply chains pre-COVID was segregated into the global players with the means and technology for seamless movement of goods and the small players who relied on fragmented and frugal logistics to offer effective and entrepreneurial logistics services. The future scenarios therefore raise the question – "*What* role will supply chain businesses assume in the future *if* retail product and service offerings are integrated and modulated simultaneously across local, communal, regional and global markets?" Supply chain businesses would transform into the creators of a global blueprint depicting the nexus of trade from global to last mile connectivity. Compared to the existing role as transport solution providers of goods and services, supply chain businesses would assume the role of the chief architects conceptualising, designing and implementing the movement of goods in a seamless or modular way, depending on which scenario unfolds in the future. They would work closely with retail tech providers to develop modular technologies to enable this transition.

4.6.4 Strategic Reframing for Retail Policy Makers

In order to be successful in the future scenarios proposed here, retailers would have to operate as collaborators within a mutually cooperative ecosystem. For example, local and regional communities would require financial and infrastructural support to operate as self-sufficient and sustainable retail hubs, especially "It's my Life" and "Heal the World" scenarios. Retailers irrespective of size would need access to technology to operate across multifarious retail channels in Robot Rocks scenario. Retail policy makers have thus far played the role of competition regulators, enforcers of rules to enable sustainable practices, protectors of mass voters employed in the unregulated mom and pop and small-medium retail sector. These scenarios therefore raise the question "*What* role will present day regulators assume in the future *if* retailers needed to operate in mutually collaborative ecosystem irrespective of their size?" To enable retailers to thrive in these scenarios, regulators would need to transform into rainmakers providing financial and regulatory incentives for large and small/medium retailers to share infrastructure, customer information and adopt technology and sustainable practices. Interventions from policy makers will continue to be fundamental in reshaping and nurturing the recovery of a retail sector in a post-COVID world. In short, strategic reframing would entail retail policy makers transforming themselves from regulators to rainmakers of the retail sector.

4.7 CONCLUSION

In this chapter, the author proposes three plausible, challenging and relevant retail scenarios for a post-COVID world which provide future safe spaces within which retailers could consider alternative strategies to develop resiliency from future macro environmental disruptions and shocks. The three future scenarios pertaining to a hyper-tech world, a green world and an insular world raise some counterintuitive "what-if" questions for retail industry stakeholders. Answering these questions challenge the existing roles of four critical stakeholders of the retail sector and highlight the transformative roles they would need to undertake in the future. First, these transformations entail retail real estate owners assuming the role of collaborative creator of retail experiences with retail brands. Second, for retail policy makers the transformations require them to assume the role of rainmakers for the retail sector instead of regulators. Third, supply chain businesses require transforming from transport providers to architects of integrated, seamless and modular collaborators for retailers irrespective of size and location. Fourth, retail brands are required to transform into sustainability educators and innovators. The author concludes that the common narrative in all the transformative roles is collaboration. Such collaborations can create a mutually beneficial nexus of new relationships between retail industry stakeholders, irrespective of their size and financial prowess, which in turn can create a stronger foundation and provide greater resiliency to the retail sector in a post-COVID world.

REFERENCES

Bobbitt, P. (2002) *The Shield of Achilles: War, Peace and the Course of History*, Penguin, London.

McRae, H. (1995) *The World in 2020: Power, Culture and Prosperity. A Vision of the Future*, Harper Collins, London.

Mukherjee, M. (2020) "Could scenario planning have saved Robinsons" Business Times, Singapore, Monday 9th November 2020. Available from https://www.businesstimes.com.sg/opinion/could-scenario-planning-have-saved-robinsons [Last Accessed 4th June,2021]

Mukherjee, M, Ramirez, R and Cuthbertson, R. (2020). "Strategic reframing as a multi-level process enabled with scenario research", *Long Range Planning* 53.5. pp 2–19.

Niblett, R. (2020) "The end of globalisation as we knew it" in *Foreign Policy, March 2020 'How the world will look after the Coronavirus pandemic'* Available from https://foreignpolicy.com/2020/03/20/world-order-after-coroanvirus-pandemic/ [Last Accessed 4th June 2021]

Powell, A. (2020). "What will the new post-pandemic normal look like?" *Harvard Gazette*, November 2020.

Ramírez, R. Mukherjee, M, Vezzoli, S. & Kramer, A.M. 2015. "Scenarios as a scholarly method to produce interesting research", *Futures*, 71:70–87.

Sanchez, P. (2020) "Opinion: A multilateral agenda for a strong, resilient, green and inclusive recovery from COVID-19" Available from https://www.oecd.org/newsroom/a-multilateral-agenda-for-a-strong-resilient-green-and-inclusive-recovery-from-covid-19-opinion-article-by-pedro-sanchez-and-angel-gurria.htm [Last Accessed 4th June 2021]

Schoemaker, P. J.H. (1995). "Scenario planning: a tool for strategic thinking", *Sloan Management Review*, 36 (2):25–50.

Schwartz, P. *et al.* (1999). *The Long Boom: A Vision for the Coming Age of Prosperity,* Perseus, Boulder.

Treadgold, A. and Reynolds, J (2020). *Navigating the New Retail Landscape: A Guide for Business Leaders.* Oxford University Press.

Van der Heijden, K. (2005). *Scenarios: The Art of Strategic Conversation,* 2nd Edition, Wiley, Chichester.

Wack, P. (1985). "Scenarios: Uncharted waters ahead", *Harvard Business Review,* 63(5): 72–79.

Walt, S.M. (2020) "A world less open, prosperous and free" in *Foreign Policy (March 2020) 'How the world will look after the Coronavirus pandemic'* Available from https://foreignpolicy.com/2020/03/20/world-order-after-coroanvirus-pandemic/ [Last Accessed 4th June 2021]

OTHER RESOURCES (DATABASES, WEBSITES AND NEWSPAPERS)

Accenture (2020) 'COVID-19: Knowing how consumer trends impact CPGs' Available from https://www.accenture.com/us-en/insights/consumer-goods-services/coronavirus-cpg-consumer-needs [Last accessed 17th January 2021]

BBC (2020) 'Corona virus: Will our day to day ever be the same?' Available from https://www.bbc.com/worklife/article/20201109-coronavirus-how-cities-travel-and-family-life-will-change [Last accessed 13th January 2021]

Economist (September 2020) 'The pandemic is plunging millions back into extreme poverty' Available from https://www.economist.com/international/2020/09/26/the-pandemic-is-plunging-millions-back-into-extreme-poverty [Last accessed 17th January 2021]

European Strategy and Policy Analysis System ESPAS (2015) "2030 Global Trends to 2030: Can the EU meet the challenges ahead?" Available from https://espas.secure.europarl.europa.eu/orbis/sites/default/files/espas_files/about/espas-report-2015.pdf [Last Accessed 4th January, 2021]

Foreign Policy (March 2020) 'How the world will look after the Coronavirus pandemic' Available from https://foreignpolicy.com/2020/03/20/world-order-after-coroanvirus-pandemic/ [Last Accessed 4th June 2021]

International Crisis Group (March 2020) 'COVID-19 and Conflict: Seven Trends to Watch' Available from https://www.crisisgroup.org/global/sb4-covid-19-and-conflict-seven-trends-watch [Last accessed 4th June 2021]

KPMG (July 2020) 'The impact of COVID-19: A global to local overview' Available from https://assets.kpmg/content/dam/kpmg/sg/pdf/2020/08/The-Impact-of-COVID-19.pdf [Last accessed 12th January 2021]

OECD (December 2020) 'The Paris Agreement 5 years on: Taking Stock and looking forward' Available from https://read.oecd-ilibrary.org/view/?ref=1059_1059337-wv6hvoqytu&title=The-Paris-Agreement-5-years-on-Taking-stock-and-looking-forward [Last accessed 11th January 2020]

The Guardian (March 2020) 'Power, equality, nationalism: how the pandemic will reshape the world Available from https://www.theguardian.com/world/2020/mar/28/power-equality-nationalism-how-the-pandemic-will-reshape-the-world [Last accessed 12th January 2021]

World Economic Forum (February 2020) 'This is what we can really do about climate change. Says new report' Available from https://www.weforum.org/agenda/2020/02/tackling-climate-change-actions-report/ [Last accessed 11th January 2021]

II

Process

The second section investigates how workplace processes have experienced changes enabled by technology. At the same time, digital transformation took place in different sectors differently as more activities moved online, and the definition and constraints of space and time warped. Digital transformation can also be a double-edged sword, as we witnessed increases in cybercrimes.

CHAPTER 5

Transformation Roadmap

Pivoting and the Emerging Trends in a Post-Pandemic World

Christopher Warren

Strategy & Consulting, Accenture, Singapore

CONTENTS

DOI: 10.1201/9781003148715-5

5.1 INTRODUCTION

The pandemic has created an unprecedented level of disruption for businesses and society. There are different industries that have been hit more severely than others. The aviation and travel industry has seen up to 90% reduction in customers and traffic, becoming dependent on government support to outlast the pandemic. While industries that deliver products and services via technology platforms have seen an increase in the sales and subscriptions. There has also been an increase in the sale of digital collaboration tools within more traditional organisations to minimise impact of remote working to business operations. Many employees, whose roles have been displaced, are looking to re-skill to participate in the post-pandemic economy.

There is an increased reliance on and acceleration of technology, digital channels and demand for people with technology skills to contribute in the post-pandemic economy. Organisations are increasing the speed of digital adoption and transformation in order to pivot their business to operate in this new economy. Educational institutions are expanding and strengthening their technology faculties to address the demand for higher-quality education. Governments are providing stimulus to transition displaced workers towards digital and emerging technology.

Technology had a major part to play in dealing with the pandemic and will continue to do so in the post-pandemic future. Ultimately, the technology roadmap will be defined by the people and organisations who can pivot quickly in the changing environment and define a journey forward with the technology at the forefront.

This chapter will outline what differentiates successful organisations that are able to pivot during the pandemic. This will lead to the three major pivots that organisations have/are experiencing as part of the pandemic situation. Finally, it will cover the emerging trends that will shape the new normal for organisations in the post-pandemic world. The intention is to provide a playbook for future black swan events that will need organisations and individuals to pivot (Figure 5.1).

FIGURE 5.1 Overview of chapter.

5.2 CHARACTERISTICS THAT PREDICT ABILITY TO PIVOT SUCCESSFULLY

The ability to pivot is essential for survival in any organisation. Preparedness to pivot comes primarily from three key sources: (1) People and culture; (2) investment in technology; and (3) being data-led. The pandemic presented a unique challenge as a black swan event that is truly global. For multinational organisations there was no 'one size fits all' solution as sovereign nations took different pandemic response approaches.

The first indicator of the ability to successfully pivot is the people and cultural norms that influence the ability to change and adoption of technology. Take the example of Singapore and the release of the Trace Together[1] initiative to help contact tracing efforts. This is the application and adoption of new technology at speed to overcome a crisis. Bluetooth technology embedded in the token records the interactions with other tokens held by other individuals. If an individual is infected with the virus, the data in the token allows for immediate contact tracing.

However, in a Western country the roll-out has been slow and encountered adoption resistance. Asia and emerging economies adopt technology faster because the technology provides a service that they did not have access to in the past. These countries are leapfrogging developed Western countries. Culturally, developed countries have a relatively good standard of living and with that comes concerns around privacy, personal data and scepticism of how their data will be used when technology is applied.

The second indicator is the organisation's approach to digital and technology investment. Organisations that have been proactive in technology investment have had a larger advantage over organisations that have lapses in their technology investment[2]. A delayed investment in technology creates a technology debt (Tech Debt) that will eventually need to be repaid with interest in order to recover and be on par with competitors that have invested in technology.

An example of a recovery from a technical debt can be seen in the financial services industry. Large portions of their operations still rely on mainframe systems that were developed over 30 plus years ago. The accrued technical debt has been significant. These systems are core to banking operations and there is a significant risk in making upgrades, moving to new facilities and interfacing with modern platforms. Most financial institutions with this technical debt develop a transformation roadmap to 'digitally decouple' the business services from the technical services. This reduces the reliance on legacy mainframes while incrementally providing modern technology services in support of more flexible customer services at the front end. It is only in recent years that digital banks begun to reimagine the future of banking without the technical debt of a mainframe.

Finally, when an organisation becomes data-led it requires the coordinated investment in the previous two characteristics. Investing solely on people or technology is not going to yield the value that the coordinated investment in both creates – an environment where there is an accepted, secured fact base and a culture and governance applied to decision making. A data-led organisation is able to see the real-time business impacts of various situations.

Once an organisation is data-led, it will have a wealth of data for people to rally around. When everyone in the organisation is able to securely review the fact base, it allows everyone to participate in the solution process. This allows process, governance and technology to work together to maximise the collective resources of the organisation and chart a path forward. Digital simulations of options provide insights to help leaders and boards decide on the best path forward. This is critical learning from the first pivot many organisations experienced as part of the pandemic response.

The organisations that have been able to survive have gone through three major pivots:

1. Emergency pivot

2. Operational pivot

3. New normal pivot

5.3 THE EMERGENCY PIVOT

The emergency pivot is the first initial triage that this pandemic forced organisations to address. The goal is to conduct a stock take of resources available, stop unnecessary spending and build up liquidity as a precaution to the depressed economic environment. The faster organisations have awareness that a crisis was underway, the more prepared they are to respond and pivot. Preparedness does not prevent the disruption but it does provide the organisation time and options to respond in the best way possible.

As an example, when the pandemic first hit, many countries established lockdowns forcing employees to work from home and remotely. A large multinational organisation with a history of acquisitions was attempting to understand the technology assets (e.g. laptops, desktops, monitors etc.) available to deploy them to employees working from home. However, due to the historical mergers there was no up-to-date, single source of truth regarding the availability of assets. Due to the presence of multiple legacy systems, there was no single source of truth to identify the location of assets. Lapses in updating records in multiple systems and minimal audit trails were limiting the ability of the organisation to respond to the crisis. The company needed to quickly consolidate, audit and deploy technology assets as a requirement to enable remote working arrangements. These activities require time, effort and come with an opportunity cost of supporting clients and customers. This severely restricts an organisation's ability to generate sales and revenue at a critical time, as employees are unable to access the tools and information required to execute their roles.

Ideally, if the organisation had invested in their enterprise systems, controls and reporting, the emergency pivot would have been conducted thoughtfully and calmly like any normal board or executive meeting – discussing the state of the broader economic environment, reviewing how operations are executing and delivering value to customers to generate revenue. The emergency pivot is vital for organisations to collectively align on the state of business and have all relevant information available to make decisions on how to respond.

5.4 THE OPERATIONAL PIVOT

The operational pivot is focused on executing as quickly as possible the capabilities that allow critical business operations to continue. This includes contact tracing, employee safety solutions and accelerating digital programs that allow operations in the pandemic such as e-commerce and digital presence.

From large multinational companies to local restaurants, all are pivoting with technology utilisation to access customers in new channels, reconfigure supply chains and reduce costs.

At the onset of the pandemic, a number of large, multinational organisations needed to roll out collaboration tools to allow their employees to operate remotely. As this technology was now considered critical infrastructure to the organisations, the investment decisions were expedited. This allowed for the implementation and deployment of collaboration tools to be conducted at greater speed and with full support.

For one organisation, the deployment of a collaboration platform was achieved within two weeks. This included not only the technical deployment but the change management and support to help employees maximise the value of the new tools available to fulfil their roles.

There are also many stories from small businesses that have implemented digital technologies to pivot during the pandemic. Local restaurants have made changes to their businesses model, moving from dine-in to take-away by leveraging eco-system partnerships (e.g. Grabfood, Uber Eats, etc.). This allows access to customers in new channels allowing revenues to continue when there are social distancing restrictions. This has also expanded as restrictions eased, as local restaurants are implementing digital menus and cashless payments to minimise opportunities of virus spread.

The operational pivot is a significant change for organisations that needs to be implemented at speed in order to survive. There is a level of urgency, confusion and disruption that makes this pivot challenging for many organisations. However, if the people, culture, past technology investment and focus on data for insights exist, the ability to execute in this pivot becomes much easier.

5.5 THE NEW NORMAL PIVOT

The new normal pivot, which many organisations are experiencing, is defining and creating a new business baseline with a clear technology roadmap to help the organisation thrive in the post-pandemic world. The major difference between this phase and the previous is that organisations cannot continue to operate on adrenaline. Organisations will stabilise and have process and plans in place for continued business operations. At some point in the future there may be a significant uplift in performance and results from broader economic improvement; however, the current status is to continue to deliver and accept the current situation as the basis for operating in the near term.

The new normal is defined by organisations returning to a state of operation that looks at short-, medium- and long-term planning to grow revenue, customers and profit. This leads to the next section of the chapter focusing on the emerging trends that will shape the new normal for organisations.

5.6 EMERGING TRENDS SHAPING THE NEW NORMAL

As organisations move into a new normal, there are a numerous emerging trends that will shape the future. There will be significant changes in the short to medium term to the role of technology within organisations. This is driven by the rapid and often forced acceleration of digital technologies caused by the pandemic response. In the near term this will involve investment in:

1. The Engine – Cloud Computing

2. The Fuel – Data and Internet of Things (IoT)

3. The Performance – Machine Learning (ML) & Artificial Intelligence (AI)

All three of these technology trends require a holistic organisation wide transformation to maximise benefits. Implementing these technologies requires reimagining how all areas and functions of and organisation operate, interconnect, and add value to the whole.

5.7 THE CLOUD ENGINE – DIGITAL SCALE AND AGILITY

Cloud is the new engine for organisations to take full advantage of the benefits of technology at speed and scale for business. In its simplest form, Cloud is the umbrella term that describes the industry and business of hyper-specialisation in technology infrastructure and network services. This model allows organisations to outsource at scale, creating fewer barriers for businesses to leverage technology. It removes the high cost of entry for leading edge technology services.

Cloud service providers (CSPs) have invested heavily in building high performance data centres and services in multiple locations around the world, allowing organisations to leverage storage and compute platforms that are scaleable based on usage and have a number of business services that can additionally optimise technology infrastructure. When signing up for these services, it moves a capital expenditure (capex) to an operational expenditure (opex), reducing the historically high barriers of entry to high quality and availability services.

There are many different cloud options available for organisations. Figure 5.2 outlines the high-level differences with the different types of cloud available. The decision on which solution to leverage can be complex. Business strategy, product strategy, regulation, risk, security, talent and cost are all important considerations when making the choice.

As Cloud is moving into the mainstream for organisations, there are many challenges and misconceptions that need to be learned. First misconception is that Cloud is cheaper. This is not necessarily true and

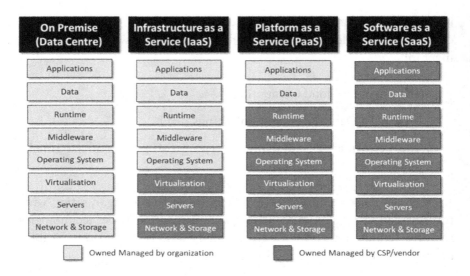

On Premise (Data Centre)	Infrastructure as a Service (IaaS)	Platform as a Service (PaaS)	Software as a Service (SaaS)
Applications	Applications	Applications	Applications
Data	Data	Data	Data
Runtime	Runtime	Runtime	Runtime
Middleware	Middleware	Middleware	Middleware
Operating System	Operating System	Operating System	Operating System
Virtualisation	Virtualisation	Virtualisation	Virtualisation
Servers	Servers	Servers	Servers
Network & Storage	Network & Storage	Network & Storage	Network & Storage

☐ Owned Managed by organization   ■ Owned Managed by CSP/vendor

FIGURE 5.2 Overview of types of Cloud options available.

organisations need to fully understand the capabilities of Cloud and how they plan to embed them in the organisation.

As an example, a retail business during the pandemic was looking at improving their e-commerce channel to engage with customers digitally when their physical stores were closed due to lockdowns. The topic of Cloud was raised as a model to accelerate this journey. The client shared that they had moved some of their workloads to the Cloud; however, despite the possible benefits of scalability and elasticity, the infrastructure costs associated with workloads was increasing. This was because the client had employed a 'lift and shift' approach of moving applications and systems to the Cloud environment without first optimising them for the underlying technology. They did not leverage the built in capabilities that CSPs offer as part of the Cloud.

A simple analogical explanation would be the client moved a box from their own storage facility to someone else's storage facility. However, they did not take the time to unpack the box to remove items they no longer needed or to take advantage of services that were previously unavailable in their facility.

Accepting and deploying Cloud triggers a fundamental change within organisations, which they will need to adopt in order to take full advantage of the services and compete in the modern digital environment.

As Cloud computing is adopted at greater speed and scale, organisations will ultimately adopt a multi-cloud and hybrid cloud environment. This will involve multiple CSPs being deployed in a multitude of configurations (IaaS to SaaS) based on the specific business functional needs.

There will be numerous large partnerships and transformation deals that will gain significant attention in the market. These will be where three key parties will come together at scale to deliver the benefits of cloud across an entire enterprise. The three parties will be (1) Client organisation with sufficient workloads to transition to the cloud at scale; (2) CSPs; and (3) large systems integrator (SI)/consulting partner to support execution. The value of the SI/consulting partner is essential given the large disruption that will occur within the enterprise. Client organisations will need to operate and transform in parallel in order to deliver the value of the investment case. These deals will be very large and span multiple years. They will require strong leadership and commitment from all parties.

In the long term, organisations will implement and require the capabilities to manage multiple cloud deployment options. This is referred to as a 'multi-cloud' or 'hybrid cloud' delivery model. The various cloud deployments could be by business function or as a holistic infrastructure transformation investment. Ultimately, they will all need to be in alignment with organisational wide architecture, design and security principles. As the business operating environment organically expands, the technology leadership will be challenged with maintaining security, compliance, and governance. This will require investment in Operational Control Centres to actively provide security assurance monitoring, consistent deployment of cloud services and threat-based modelling. These operational control centres, either physical or virtual, will allow technology leadership to view, monitor and respond to the complex landscape supporting the business. They will contain cloud brokerage functions to manage multiple CSPs, allowing workloads to be migrated from one cloud to another based on business needs or cost; along with tools to govern and maintain security and compliance. Technology leaders and departments will require these tools and solutions as they will be required to focus their support on business outcomes. The underlying complexity of Cloud technology should not distract or hinder the strategic business objectives of the organisation. Technology will be intertwined with all products and services making every organisation a digital/technology business.

5.8 THE DATA FUEL – INTERNET OF THINGS (IoT) AND INDUSTRY X.0

Once organisations have been able to establish a cloud foundation, organisations need to fuel the Cloud with data. Data is the currency of post-pandemic world and the Cloud provides a scaleable on-demand solution to build information and insights.

The largest growth area for data is the convergence of Information Technology (IT) and Operational Technology (OT) in what many are calling the Forth Industrial Revolution or Industry X.0. This is characterised by the application of technology to operational functions of an organisation. This typically takes the form of robotics, sensors and IoT devices capturing data and conducting analysis/decisions at point of capture (edge computing) or ingested into a central control centre.

Edge computing will be further enabled by roll out of 5G technology. The increased network speed and capacity will allow for larger amounts of data to be transmitted. This will improve the responsiveness of many solutions allowing for mission critical systems to operate more effectively. This will include autonomous vehicles, robotics and operations centres (e.g. airports, manufacturing plants).

As an example, an engineering company looking to optimise the value from their Maintenance, Repair & Operations (MRO) Function with a Smart Control Centre. The smart control centre would look to ingest as much data as possible from existing systems and IoT devices in the field to inform management on the holistic view of the organisation. It would also identify end to end where there is missed value and opportunity and provide visibility into the delivery and ultimately revenue associated with completing task on time.

This organisation is no different in that it plays to its strengths and approached technology and innovation with an engineering mindset, identifying the problem and work through to develop the best solution possible for the problem. This is a highly effective approach to resolve issues; however, it does present challenges when you take step back to look at the broader objective.

Building a smart control centre to provide end-to-end visibility across the value chain requires multiple systems and devices to be connected to a common fabric. The challenge with the engineering approach is that without a common strategy, architecture and vision, each of the sub-solutions require significant rework to integrate to the centralised control centre. In the environment there were some truly innovative solutions that leveraged great technology; however, without a common strategy, the ability to integrate the patchwork of solutions required significant effort. Once a clear vision and architecture was defined to frame the current solutions and any future solutions, the ability to create the pool of data to fuel the smart control centre was possible.

Defining a clear vision, supported by strong architecture and data strategy builds great innovative solutions. In some cases, it takes time to identify through the complexity that individual projects and initiatives are intertwined with dependencies. Once this opportunity has been identified, developing a cohesive strategy and taking advantage of the opportunity by 'Measure twice, cut once' would yield significant benefits for the organisation.

5.9 THE AUTOMATION PERFORMANCE – MACHINE LEARNING & ARTIFICIAL INTELLIGENCE

With data fuelling the Cloud engine it creates an environment for the rapid growth of artificial intelligence (AI), machine learning (ML) and automation at scale. This growth will unlock large performance improvements for organisations that in turn allow employees to focus on higher value and innovation activities. In order to grow these technologies, large amounts of data need to be processed at speed to train models and conduct scenarios against historical data. This would deliver a new and improved level of performance for the organisation allowing them to maximise the technology stack to achieve high efficiency, insights and value.

As an example, a 1,000 store strong retail client in South East Asia was looking to develop a platform with AI and ML capabilities to assist in making decisions on store locations. Leveraging a subset of historical sales and revenue data, coupled with external data (e.g. population, geolocation, competitor locations etc.) they were able to produce a model that, when tested against unseen historical data, proved to be more than 80% accurate in predicting future sales, revenue and cannibalisation of nearby stores. The value generated by this platform allowed the client to save time on store location analysis, optimise property investment holdings and maximise the ROI on future expansion.

This example highlights that analysis, recommendations and decisions can be made faster and more reliably than a human in the role. This will lead to the most important challenge of governance and providing transparency in the automation. Governance seeks to address who has the authority and controls in managing and changing automation. Having transparency of the automation give customers, partners and suppliers the confidence in how decisions and processes are executed. In many cases, the automation will initially be perceived as an opportunity to reduce labour costs. This should not be the approach that organisations take if they are to leverage the full value of automation.

Deploying and managing an automated workforce creates a number of efficiencies.

1. Automation can operate in environments that are hazardous for human workers. As an example, autonomous vehicles can be operated remotely and without a physical driver reducing the risks to employees.

2. Automation can operate in environments where conditions can be optimised for cost. As an example, warehouses that are operated with process automation and robotic pickers are able to complete their roles in the dark without lighting. This provides energy cost savings and environmental benefits.

3. Automated workforce do not require healthcare and do not require motivation to complete their designed tasks. Therefore reducing the management overheads associated with commoditised and repeatable tasks.

Organisations with an automated workforce create reliability, efficiency and cost benefits to the bottom line that disrupt the role of a human within an organisation. The post-pandemic future employees will require a higher level of skill sets to be productive. The skills will be focused more on the monitoring and repairing of automation. In addition, more people will need to be employed to analyse the outputs and craft insights to allow business growth. This includes the more creative components such as designing new cost reduction initiatives, revenue channels, and merger and acquisition (M&A) opportunities. Automation assists in analysing at speed, however architecting and designing the action will require imagination, innovation and perspectives that automation is unable to currently provide.

The combination of these investments will change the role of technology within organisations. Technology will have the ability to operate as a front of house function and bring forth new operating models, ways of working and leadership.

5.10 CENTRE STAGE – THE SHIFTING ROLE OF TECHNOLOGY

With the deployment of Cloud, data and AI, the future technology environment will be nebulous. Technology as a standalone capability will disappear as it will become ubiquitous with business itself. Talent in the modern economy will embed technology into the traditional industry and profession. It will be enabled by a collection of high-performing autonomous teams. This change is already beginning in many organisations. However, there are significant challenges as traditional leadership, processes and controls are not suitable for this new work environment. The primary barrier to this change is culture and mindset of the people within the organisation. It is difficult for everyone to take this leap, and even more difficult for people who have found success in the traditional models. If the current way of working is delivering results and rewards for individuals, they have little incentive to change.

The first and most obvious change is that technology will become front and centre to the value generated by organisations, fully embracing the concept that every company is a technology company[3]. As an example, recently published articles focus on how the specific technology behind visual effects (VFX) is disrupting the development of movies and television[4]. Traditionally, VFX was part of the post-production process that only occurred after editing. The VFX team would receive the near-finished product and apply post-production effects such as removal of green screens, light correction and any other effects that improve the immersive experience of the film. With technology advancements, tools and agile ways of working, the VFX team can now be an active participant in the filming process. A team of digital artists is now able to work in state-of-the-art sets that are enclosed by high-resolution LCD panels[5]. In advance of filming or in near real time they can make updates to the digital set. Working closely with the set designers, sound operators, actors and directors brings greater agility, speed and value to the final product. The work environment has moved for a back-office function to a front-line value generator (Figure 5.3).

This example is a brief illustration that also helps other organisations identify the challenges that will needed to be overcome in order to achieve the agility, speed and value. Moving technology from a back office to a front office function creates large disruption in the organisation. People are now required to interact, accommodate and collaborate with individuals who were previously not even visible, thus creating a new understanding and appreciation of other skills, defining new ways of working,

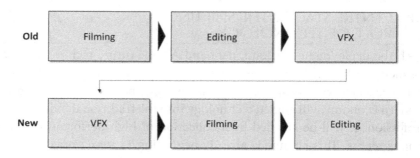

FIGURE 5.3 The changing role of VFX in media (film and television).

processes and governance. This does create challenges for the organisation as anxiety and disruption in the workforce dilutes value generation in the short term.

5.11 THE NEW MO – THE END OF MODUS OPERANDI

As technology moves to the forefront of organisations there will be significant changes to the ways of working and operating models to deliver results and outcomes. The concept of developing standard processes that can be 'set and forget' will not be sustainable. The rapid change in technology, business models and complexity will require a significant change in mindset. The effort and resources required to maintain detailed up-to-date processes and procedures will not yield the value that may have existed in the past.

Technology skills and talent will become decentralised across the organisation. In the past, teams were often located together with some large business and technology functions taking up floors of office space. In the new normal, technology talent will be distributed in small teams across multiple functions in different locations. Embedding technology at the forefront of business functions, for example finance, supply chain, customer service etc., will require multiple individuals in teams to execute business strategy as close to the customer as possible.

This creates significant disruption in terms of collaborating, cross pollinating ideas and driving innovation across projects and teams. The focus shifts from a structure hierarchy and centralised knowledge base to an environment of multiple high-performing, multidisciplinary, autonomous teams.

Take for example the consulting and advisory industry. In order to deliver value to clients, the goal has been to establish combinations of high-performing individuals from across capabilities to solve specific client challenges. The consulting industry has continually struggled to solve this challenge with many firms continually switching their organisational structure from capability/function focus (e.g. technology, supply chain etc.) to industry focused (e.g. travel, government and public sector, financial services etc.) based on the leadership perspectives at the time.

Newly developed and deployed operating models aim to short circuit this back and forth. In doing so, the focus has shifted towards specific client needs, leveraging the skills and experience of the individual people across a series of networks regardless of capability or industry alignment. This change results in a greater focus on client account leadership, enterprise architecture/holistic vision for clients and development of leadership and management skills that support the building of high-performance teams.

However, there are challenges, as traditional models rewarded leaders that built and scaled practices of individuals with homogeneous skills and experience. This created chiefdoms and land grabs creating an internally focused firm, rather than a client-centric firm. In developing new operating models for consulting and advisory, the traditional practice development approach needs to be dissolved while retaining a sense of community and belonging for staff. It is a challenging transition for many; however, the objectives and benefit from this new way of working deliver greater value to clients.

With a change in operating model a change in leadership style is also required. Traditional command and control models are not as effective in a network environment.

5.12 STEWARDS AND STALWARTS – THE FUTURE TECHNOLOGY LEADERS

An organisation that implements emerging technologies, ways of working and governance requires a leadership skill set and talent that does currently exist at scale. The role and skills of Chief Information Officers (CIOs), Chief Technology Officers (CTOs) and Enterprise Architects will be critical to deliver value in the post-pandemic era.

Technology and new operating models are providing the ability for leaders to have real-time or near real-time access to information to make decisions. The uninterrupted stream of information coupled with real-time execution capability moves businesses away from scheduled cycles towards a continuous journey. In the past, an organisation may undertake an activity to craft a 3 or 5 year technology strategy in support of the business. Operating an organisation on these periodic strategy cycles is no longer sustainable, especially when we are faced with external events such as a pandemic that can turn that strategy to paper weight within seconds. The new normal creates an environment where leaders are stewards supporting the broader and continuous organisational journey. Experience is still essential for leaders; however, the role of a leader in this environment is significantly different from a traditional leader. Key differences include leading by influence rather than direct control and bringing together multidisciplinary and diverse teams for specific tasks and outcomes.

One negative observation and challenge of some leaders during the pandemic has been the increase in 'spreadsheet leadership'. This is where leaders, in the absence of regular in person discussions, have focused their attention on what can be measured in reports and data. This is a natural response as most leaders have not had the experience of leading a remote team. Making decisions and leading teams with a heavy bias upon formal reporting does not necessarily empower, motivate and direct teams in the right ways. Leveraging data for decision making is key and part of the importance of driving a data-led organisation. However, it needs to be coupled with engagement. Data and reporting needs to be understood, have qualified sources and be relevant. If the data is not accurate it can add to the stresses of the organisation. Post-pandemic leaders will be able to differentiate themselves and grow by bringing the best of both metrics and empathy to the workplace.

Many individuals have faced some level of struggle with working remotely – the endless video calls and the absence of water cooler discussions. Individuals have been unable to connect impromptu, and that also applied to leaders. Leaders that need to collaborate on joint decisions have been constrained as the time is no longer available. A simple example is a procurement process. A procurement panel is established to review proposals and select a vendor to execute and deliver on a plan. In the past, leaders were able to connect briefly sharing their thoughts as they move from one meeting to another. In a scenario when all leaders are working remotely, they each must find specific and dedicated time to connect and share their thoughts. The impact of this can lead to mixed messages being delivered to the procurement team, poor communication with vendors and extended procurement cycle. This ultimately leads to delays in executing delivery, business plans and results. The pandemic created a huge

FIGURE 5.4 Characteristics of future technology leaders.

change impact on the ways of working that will need to be addressed in the future as remote working.

Take the role of the CIO, CTO or Chief Digital Officer (CDO). These roles will still be unique, as their years of experience in technology will still have considerable value. However, without direct control over the IT talent they will need to become more of a mentor, teacher and steward of the technology vision to influence and guide the business rather than dictate and control. It will require some of the following characteristics (Figure 5.4):

- Have a more proactive approach to engage across the organisation, speaking more about ideas, implications and ethics; especially with regards to AI.

- Have a role to play in educating other executives and the board on the value and implications of technology in relation to the business.

- Active engagement up and down the supply chain so as to identify opportunities for digital collaboration and value generation across the ecosystem.

- Active engagement with strategic partners and suppliers to co-create the future value and expectations.

- Possessing an open mindset that questions technology proposals, rather than correcting them to technology and architecture policies. This will support broader organisational learning and growth.

The role of a CIO, CTO and CDO will continually evolve with the business environment. Taking time to reflect on the skills, qualities and purpose of the future is not only important for current but also for aspiring technology executives, but also for the broader business leaders. Organisations need to regularly engage in a dialogue with their digital leaders about the needs and value they wish to unlock with the use of technology. It is this conversation that will lead to growth and resilience for businesses to actively engage in the post-pandemic economy.

The enterprise architect is also a role that will need to be empowered and supported in the post-pandemic work. This role has historically come with technology focused connotations; however this is a fallacy. A true enterprise architect has the visibility to develop logical models that cover all elements of an organisation. This big picture perspective when articulated correctly provides all stakeholders a common map of the organisation.

The position of technology, ways of working and leadership disruptions are occurring as emerging technologies are adopted. No longer will the technology talent be consolidated within the technology department and it will become more fragmented as hyper-specialisation continues in different technology fields. Supply chain, finance, marketing, legal, research & development, manufacturing etc. will all be supported and run by teams of people that have both the functional knowledge and technical knowledge. As organisations understand these changes, they will need to continually adapt to the emerging technology such as Cloud, data and insights and AI.

5.13 LONGER TERM – CONVERGENCE AND COLLABORATION ACROSS INDUSTRIES

In the longer term, the results of the pandemic will drive deeper conversations across the societal and economic ecosystem. Specifically, the intersection of technology, education, business and government will generate opportunities and collaborations that will alter economies and societies globally. The outcomes of this collaboration will lead to new regulations, compliance and potentially creating new industries.

As technology scales across organisations, it will have significant implications for the workforce and society. It will reduce the time and effort employees spend on commodity and administrative tasks. Similar to the invention of agriculture[6], automation will generate more capacity in organisations to innovate, capture market share and create value. To generate additional value, employees will need to attain a higher-level skill set and aptitude to participate in the workforce. They will need to understand how the technology operates and is architected in order to manage the risks and dependencies of changes that are made in the environment (Figure 5.5). This is also true of other major black swan events. It causes society and

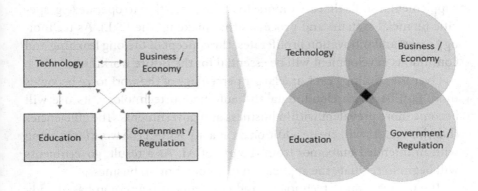

FIGURE 5.5   Changing ecosystem relationships and industry convergence.

economies to re-evaluate the relationships and build stronger partnerships and institutions to protect against such events in the future.

Process and transaction-based roles will slowly become automated where possible, with employee skills moving to analysis, insight and innovation. Skills will consolidate around two key areas:

1. Deep technical engineering skills required to build and maintain Cloud, data and automation assets; and

2. Critical thinking, communication, leadership and innovation to interpret, make decisions and execute on new value opportunities.

This places a greater level of importance on organisations to build the workforce of the future. The challenge for companies is to answer the question of how much they should invest on employee development. The average tenure of full-time salaried employee (in the United States) is 4.1 years[7]. Therefore, for maximum return of investment, training should be provided as early as possible from when an employee is onboarded to an organisation. This maximises the productivity and skill gains that can be applied within the tenure of the employee.

The importance of training and education also presents opportunities for greater collaborative partnerships between educational institutions and business organisations. Collaboration between these groups can help identify the source of skill development. Educational institutions provide baseline skills in a particular topic, creating a pool of candidates for the workforce (e.g. Bachelor's in finance). Businesses need to build the supplemental skills that are unique to the organisation to operate (e.g. specific financial systems and processes that make up the P&L). As technology scales and innovation proliferates, the concept of lifelong learning and continuous development will be essential for the future workforce.

Over the coming years, as a larger percentage of the workforce becomes more familiar with cloud, data and automation technology, its use will become more prevalent within business and governments. The efficiencies that this creates will create additional capacity within the workforce along with unintended outcomes from the use of AI. As a result, governments will begin to regulate the application of automation in business.

The intersection of technology and government regulation has already begun with discussions into anti-trust[8] and discussions of 'robot taxes'[9] to balance the competitive and economic landscape. This will be a lengthy process as hyper-specialisation, interdependent components and global reach of technology creates a complex landscape to govern. Understanding the consequences of regulation on businesses and technology from ethical AI to role of social platforms content responsibilities are exponentially complex. Traditional business models, legal precedents and institutions are struggling to keep pace with global, borderless and powerful nature of the underlying technology.

As leaders in politics and civil service build capabilities and knowledge in the underlying technology new regulations and industries will develop. A trend that will appear in the long-term will be the emergence of a new Technology Audit industry. It will be similar to the establishment of the modern financial auditing industry, now dominated by the 'Big 4' auditors (PwC, EY, KPMG and Deloitte). It will most likely manifest as a requirement for organisations leveraging AI and complex technologies to be regularly audited, in addition to regulating the use or misuse of technologies. This will provide employment to more highly skilled people entering the workforce. In the same way financial services industry has created hundreds of thousands of jobs worldwide, AI and automation audit will create an entire industry of technology audit. The work within the audit field will be focused on auditing and assessing automation and AI against a set of standards and principles, highlighting to companies, regulators and governments on the outcomes resulting for the application of automation in business. This would highlight bias, fraud and compliance lapses.

The roadmap for the creation of the technology audit industry will be gradual and most likely become an extension of existing audit capabilities.

Firstly, there needs to be a critical mass of automation and AI embedded in organisations – built and operated by the first generation of AI/ML talent. Secondly, the education of government and regulators in understanding the technology implications will need to be defined, modelled and discussed. Governments and regulators will call upon the current industry leads that have developed many of the AI/ML capabilities to craft best practices on topics such as ethics, implementation considerations and regulations. This will be a long process as the implications of the regulations will impact business, education, and ripple through the existing economic structure. Once complete, this will become the catalyst to establish the industry by infusing ethics, best practice, security, governance and regulation leading to a boost to education and workforce transformation. Thirdly, educational institutions will focus heavily on coordination with governments to up-skill and re-skill the workforce to meet the newly created demand for talent to enforce the regulations, thus creating the industry. As many large organisations operate globally, there will be a need for international cooperations on some of these regulations due to the borderless digital environment.

In addition to development of technology audit industries, there will also be a consideration to the implications of a remote workforce. Consulting companies have been operating in remote environments for decades with technology systems and processes to record the working location of employees for global tax compliance. As the pandemic has accelerated the adoption of technology to allow remote working, many employees have seen an opportunity to move to locations that have a lower cost of living. If an employee can deliver the outcomes and value without being physically present, there is no need for them to be centrally located in a high cost city. This presents an interesting challenge for businesses, as it is creating a level of transparency into how they see remuneration and value to employees. Early examples are the announcements from Facebook and Slack to factor salary by location[10]. The long-term ramifications of these decisions are yet to be seen; however, it is clear for now that companies with traditionally less mobile workforces will need to understand the implications on tax reporting if there employees work remotely in an alternative location permanently. As the implications of these decisions by companies roll out, it will only be a matter of time before governments and regulators will investigate the longer-term economic impacts.

As economic, education, technology and governments converge and collaborate at scale there will be an upheaval to traditional business models.

As data driven decisions and solutions become commonplace it will create opportunities for data and information to be shared at scale. This leads to the development of smart cities and digital communities that create additional value to society. Consumer, citizen and business services will innovate based on their current service offerings to a point. New industries and platforms will be created at the aggregate level when connections identify value between these services.

The greatest challenge for the development of smart cities is the constraints created by existing business and economic models. The technology available today is able to provide citizens and consumers seamless, impactful and timely experiences. However, today these services are siloed by the organisation that provides the service. There is currently no end-to-end connectivity between services.

For example, in supporting clients develop the airport of the future, building trust and enabling collaboration between existing parties is the largest challenge in building a seamless traveller experience. The services an individual consumes in order to travel from one location to another are numerous. A traveller can book a ride on Uber to travel from home to the airport, check-in to their flight through their airline application, navigate through the terminal (security, customs etc.). They may go duty free shopping using specific store loyalty applications, before boarding their flight selecting their meal and entertainment preferences. This process is then repeated when the person lands and arrives at their destination. These processes are currently under review to incorporate public health requirements to protect countries from the speed of infections.

From a technology perspective, there are no barriers preventing the information to be aggregated to provide a seamless journey from home to destination. From checking-in luggage at home to ground transportation that arrives exactly when needed to maximise time efficiency based on traffic. The airport terminal of the future will then be enabled with contactless technology to allow travelers to board their flight without the need for physical security and customs clearance processes. Facial recognition and advanced analytics will allow security and customs services to clear individuals without the need for disruption in the customer experience flow from airport drop off to boarding at the gate. All of these services are possible today and are only constrained by investment and the business models of the existing service providers.

The airport of the future can easily be defined as a small smart city environment aggregating customer and citizen services across number of entities through the sharing and handoffs of data – government departments (e.g. customs, border patrol, transport authorities), telecommunications companies, business and infrastructure operators. Individually these entities are unable to provide the seamless impactful and timely experience across the environment, however convergence and collaboration will allow for these services to develop.

To achieve the efficiency and revenue outcomes for smart cities or airports, it requires discussion, negotiation and mediation to develop a win-win situation for all parties. This will require trusted independent third parties who understand each of the industries and businesses with the technology that can enable the services across industries. Consultancies, think tanks and standards bodies will be critical in navigating the complexities in defining how technology will enable the new normal. As example, when responding to large infrastructure tenders from government authorities, they are now beginning to refer to integrated smart city solutions. However, there is no single firm that can provide the end-to-end experience outlined by the requirements. In developing a response, a consortium of organisations is required. Existing businesses need to collaborate and negotiate in order to work out what and how they will deliver, along with the returns they expect. This requires significant effort and time to discuss between parties before they are able to provide a single cohesive response. This highlights the importance of new leadership model that requires influence and networks in order to achieve outcomes in the post-pandemic world.

The benefits of convergence and collaboration create an optimistic vision of the future, benefits of sustainability through the creation of circular economies that feed off each other. Providing digital services that are customer & citizen centric will drive product development and industry disruption. Supplying these communities will rely heavily on digital supply chains and smart manufacturing to help provide stability and security. The value generated from convergence and collaboration with appropriate regulatory oversight can provide communities and the global society with significant improvements in standards of living and environmental balance.

5.14 CONCLUSION

The pandemic is a black swan event in the ongoing journey of innovation and progress. The post-pandemic world will be dominated by those who have been able to pivot and leverage technology to survive. The challenges surrounding technology adoption, economics, regulation and workforce have grown. The pace of change will not abate in the post-pandemic world, and we must strive to make best efforts to understand, learn and progress. The challenges of complexity, hyper-specialisation and competing priorities require fundamental elements to take the journey forward:

- Technology is no longer a back-office function; we will need technology to have board and executive level exposure within organisations and ministerial or equivalent portfolio level within governments.

- We need to invest in education, and approach as a lifelong learning journey.

- We need to balance the development of hyper-specialised technology skills and holistic enterprise architecture, strategy and leadership skills as risk mitigation to the negative implications of technology.

- We will need strong multidisciplinary and diverse teams to collaborate.

- We need to empower individuals with the skills to communicate clearly and with empathy to those impacted by the disruption created by technology.

- We need accepted principles of responsibility, ownership and humility in order to participate in constructive conversation.

- We need strong institutions to govern, with the knowledge and authority to balance the economic growth, social responsibility and technological implications.

- We need to build trust in our interactions both at a personal level and between organisations.

No industry or organisation is immune to the disruption that will define the new normal. All functions within an organisation will be impacted. As more data and information highlight insights and opportunities for value creation, there will be disruption across the market. The pandemic has accelerated this change and all organisations will need to adapt and pivot in order to survive in the new normal. Collectively, members of families, communities, organisations and nations will architect the future in the post-pandemic world.

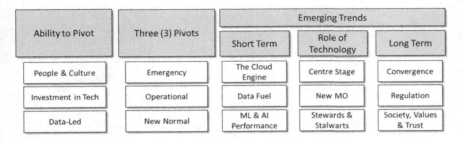

Ability to Pivot	Three (3) Pivots	Emerging Trends		
		Short Term	Role of Technology	Long Term
People & Culture	Emergency	The Cloud Engine	Centre Stage	Convergence
Investment in Tech	Operational	Data Fuel	New MO	Regulation
Data-Led	New Normal	ML & AI Performance	Stewards & Stalwarts	Society, Values & Trust

FIGURE 5.6   Summary of topics defining the future transformation roadmap.

Creating value from technology in a post-pandemic world will be achieved from bottom up innovative solutions meeting with a cohesive top down societal fabric that is imbued with the collective values, culture and leadership characteristics of those that contribute (Figure 5.6).

NOTES

1. Singapore Trace Together Initiative. https://www.tracetogether.gov.sg.
2. https://newsroom.accenture.com/news/accentures-future-systems-research-reveals-companies-that-excel-at-scaling-technology-innovation-generate-double-the-revenue-growth.htm.
3. https://www.wsj.com/articles/every-company-is-now-a-tech-company-1543901207.
4. https://www.vox.com/21529002/green-screen-mandalorian.
5. Industrial Light and Magic StageCraft solution allows for principle photography to on photo-real virtual sets that are rendered in real time and indistinguishable from physical alternatives. https://www.ilm.com/hatsrabbits/ilm-stagecraft/.
6. Guns, Germs & Steel – Jared Diamond. Highlighting the advancement of farm based societies to generate more resources than required to create capacity to innovate and exchange knowledge leading to growth and expansion for some civilisations over others.
7. US Bureau of Labour Statistics. https://www.bls.gov/news.release/tenure.nr0.htm.
8. House of Representatives report on anti-trust of Big Tech. https://judiciary.house.gov/uploadedfiles/competition_in_digital_markets.pdf.
9. Discussion on robots that take peoples jobs should pay taxes. https://qz.com/911968/bill-gates-the-robot-that-takes-your-job-should-pay-taxes/; https://www.wsj.com/articles/the-robot-tax-debate-heats-up-11578495608.
10. Facebook staff salaries to be adjusted based on the cost of living in chosen location when working remotely. https://www.ft.com/content/1c52a7a2-aa65-11ea-abfc-5d8dc4dd86f9.

CHAPTER 6

Strategic Engineering to Develop Strategies during a Crucial Period

Agostino G. Bruzzone[1] and Marina Massei[2]

[1]*University of Genoa, Genova, Italy*
[2]*University of Genoa, Genova, Italy*

CONTENTS

Looking around, sometimes, we could have the impression that decision makers, institutions and authorities, as well as major corporations don't have a so clear view about their own strategies, nor even the capability to direct their

DOI: 10.1201/9781003148715-6

resources in smart way to achieve success. We have many examples in terms of China influencing growth in Africa with respect to Western Countries (Eisenman and Kurlantzick, 2006) or failure of major successful companies such as Chrysler, Delphi, General Motors (Heracleous and Werres, 2016).

This situation looking back to past decades is not really new, but in some way, it could seem degenerating in recent times; there are potentially many factors affecting this issue, including the shrinking of the time horizon to achieve results, the volatility of the markets, the strong interconnections at global level, a large number of emerging new players, the emphasis on short-term objectives (e.g. next quarter or next political campaign), etc. Obviously during a major crisis or a crucial period, the factors turn to be even more myopic and have a high risk of failure, missing the opportunity to develop an effective roadmap to react and exit from the contingencies.

In facts the problem is even worst, considering that crisis periods correspond usually to big opportunities and missing to get these ones could correspond to a "strategic" disadvantage and a major fallback.

Therefore, we should consider that probably the strategic decision makers are required to develop new approaches to be more effective into understanding the global situation within a comprehensive environment (Bruzzone, 2018a).

Indeed, nowadays, the advances in crucial sectors such as Simulation, Modelling, Artificial Intelligence and Data Analytics are developing new impressive capabilities, especially in connection with their ability to process and use the massive quantity of data generated by company and institution digitalisation initiatives and by technological enablers such as Internet of Things (IoT) and Sensor Networks. Looking at this panorama, it is quite evident the the innovative discipline called Strategic Engineering provides a closed-loop approach to decision making based on quantitative models and algorithms and is able to guarantee major advantages in decision support.

It is evident that a major crisis further stimulates the necessity to develop strategic decision making dealing with time sensitive frameworks and multi variable problems over complex scenarios; due to these reasons Strategic Engineering is proposed as an important aspect in crucial periods (Bruzzone et al., 2019a).

In fact, Simulation Team with Genoa University is establishing multiple collaborations in this field with universities (e.g. James Cook University Singapore, Universidad Autonoma de Barcelona, University of Aix-Marseille, etc.), agencies (e.g. NATO, M&S, COE) and companies (e.g. Accenture, Ansaldo, Hitachi, Leonardo, Thales, etc.) to promote this event. Simulation team is developing Master of Science program and PhD program at international level and there are many initiatives in this field specifically addressing crisis management and future challenges.

In this chapter, the concepts and architectures of strategic engineering are presented as along with the needs and the experiences carried out on strategic engineering over a wide spectrum of application fields.

6.1 STRATEGY: WHAT DOES IT MEAN?

Let's start from a Plato's quote where he point out that: "a Strategos owns the Art to Know better what is Happening or is Likely to Happen" (Laches, 198 E, 423 B.C.). Now, in ancient Greece City States, the Strategos were the Generals and Admirals devoted to lead the military forces and it is evident that the capability to understand the current situation better than others and the ability to estimate more correctly the impacts of their own decisions on the future were fundamental to achieve final victory. This capability is obviously crucial for success in business and during crises or challenging periods; obviously, this aspect deals with the personal characteristics of a Leader, therefore it is evident that nowadays we can support a decision maker by technology, models, information as well as by a proper methodological approach in Strategic Management.

Currently, there is some kind of confusion about the meaning of the term "strategy"; many people consider "strategy" as a long-term and high-level objective, often, mostly related to planning. Therefore, this word derives from military concepts that have been investigated by major authors along centuries. A very good definition of strategy is provided by the General Anton Jomini, the Swiss Napoleon's theorist in terms of warfare: "Strategy is the art of good leading your resources" (Prècis de l'Art de la Guerre, 1838 A.D). So, it is evident from this definition that strategy is not just planning, but it includes management, successful achievement of the objectives, capability of the fixing the goals, detailed definition of desired end state, preparation of assets, reaction to situation evolution and to other player actions/reactions.

In this wide view, the strategy concept fits perfectly to the need to develop and implement solutions along a crucial period as well as the opportunity to use the new discipline of strategic engineering to support decision makers.

6.2 WHAT IS NEW IN STRATEGIC ENGINEERING?

After defining what is a Strategos and what strategy means, it is important to understand the meaning of strategic engineering and to outline its innovative nature (Bruzzone et al., 2018c).

A possible definition for this new discipline is: the combined use of Modelling, Simulation, Data Analytics and Artificial Intelligence (AI) in closed loop with real data to support the strategic decision making.

Now, despite the pretty high popularity of the abovementioned methodologies, we should be aware that they date back from quite a lot of years:

- Modelling and simulation (M&S), in terms of modern computer simulation, was introduced by gifted engineers and scientists, such as John McLeod, just after the WWII (World War II) to support development of new jets moving from subsonic to transonic and supersonic flight, as well as new rocket science, this dates back ¾ of a century (Mcleod, 1972)

- Applications of AI are even slightly older and they deal with Alan Turing development during WWII (Turing, 1948).

- Moving to data analytics, several people consider the Moving Average (one of most basic and reliable predictive algorithm) to be well known also by Romans to support logistics related to demand of amphorae of wine or tons of wheat, over two millennia ago.

Since the origins of these methodologies, we had many models and several interesting researches have been carried out to combine M&S, AI, optimisation and decision making even at strategic level (Clymer, 1993; McLeod and McLeod 1995; Bruzzone and Signorile, 1998; Piera et al., 2004; Li et al., 2017).

Based on this simplified overview, it could seem that strategic engineering is just a *buzz word* to cover well consolidated concepts already in use, but this is not correct as it results as soon as we move a little bit more on the details and we make a serious analysis on the new characteristics of this discipline.

It is true that the basic elements of strategic engineering exist since decades and that researches and applications related to their integration were conducted successfully in the past; therefore, their general architectures were relatively different in the past as well as their capabilities and diffusion for several reasons (Bruzzone, 2018b).

First of all, the capabilities of AI and M&S are outstanding with respect to the past due to the evolution of new theories, methodologies and implementation solutions (Ghahramani, 2015; Turner et al., 2016). Today there are tools and solutions to support creation of these systems that speed up the development time, reduce costs and simplify verification and validation (Lehmann and Wang, 2019). In addition, it is evident that the new available computational capabilities enabled the practical use of complex models, obtaining quickly results at reasonable costs in a much wide spectrum of applications with respect to the past (Pal et al. 2015).

Therefore, probably the major difference deals with the scenario knowledge that drastically improved for several reasons. First of all, it is evident that to properly understand the current situation of a complex system and to know how our scenario evolves, it is fundamental to have data and to be able to process them by effective algorithms and models. In reference to data, it is very well known that the concept "Garbage in Garbage Out" was introduced by Charles Babbage; indeed, along the last decades a huge effort was required to address data collection, verification, validation and certification (Williams, 1998). These data are fundamental to feed the algorithms and models; if the data are characterised by limited samples along short periods, as it was common in the past, the scenario awareness and prediction on how a situation could evolve could only offer very limited and unreliable results (Amico et al., 2000).

Nowadays, the situation is drastically changed in terms of data quantity and frequency, simply because the data collection systems are supported by the already existing and intensive digitalisation of companies and institutions as well as at individual level due to social networks and IoT (Isaksson et al., 2018; Ghani et al., 2019).

To point out these aspects, we propose a simple example: just up to 25 years ago, all major cities developed very sophisticated models of Urban Traffic obtaining even single car simulation for towns with millions of people (Ozaki et al., 1996). These models were very powerful and in use in big control centres, mostly devoted to manage congestions and crisis and they were costing millions of US dollars; therefore, few people were able to get a benefit from this solutions and their use was limited. In this sense, many models were also affected by low reliability of the predictions, from this point of view, one of the major shortfall was the fact that the data were collected by traffic monitoring systems including occasional manual counting, mechanical counters, inductive loops, etc.; all these systems were located in few sites and often subjected to discontinuity due to lack of budget and/or maintenance. As result of these elements, the data used to feed the models often had low density and reliability.

Today, we collect data in much smart way, from cameras as well as from IoT on the cars; this results in a high intensity and persistent monitoring of traffic that, obviously, allows us to control and correct models continuously, drastically improving their reliability (Wu and Wang, 2008). Obviously these systems are subjected to other vulnerabilities and problems pretty interesting respect crises, therefore in general the new approach results pretty good (Kitchin et al., 2016).

Just to complete this comparison, it is evident that in previous centralised control rooms the information was not easy to be diffused and distributed to potential users: just few electronic road signs with few words, some info by radio or call centres. Often the information was simply "traffic jam at 2 km" or "time to travel from A to B: 30" but even in these cases, considering the low reliability of data and model output, the users were soon realizing inconsistencies between the real situation and indications, while they were unable to access further valid details for better understanding the current and future situation. Today, by a simple smartphone, we access models that get such high density data to present pretty valid present and future situations, almost in real time and with minor errors. So, the user has full access to view this information in intuitive way; users quickly recognise a good reliability of such systems developing a strong trustiness on their suggestions, so the drivers follow these indications.

This is a basic example of what it changed. Therefore these achievements are not only related to the availability of large quantities of real-time data by sensors and IoT. In fact, today, the key point is on their Big Data elaboration. Indeed, the Big Data are often incomplete, unevenly distributed, inconsistent and even have plenty of errors. So, the real difference with respect to the past is that the modern techniques based on data analytics and AI are today able to access such Big Data and process them to filter, analyse and extract reliable information very quickly, and able to correct the model parameters in closed loop with machine learning to guarantee valid results from a mess of numbers.

Obviously, the computational power and the continuous huge quantity of data make it much more easy, nowadays, to obtain these good results, as compared to the limited frequency and quantity of data of the past, but it is the renewed capability to process them that guarantees us a major improvement.

However, we consider a wide point of view to understand strategic engineering. In fact, this discipline relies on different elements operating into a dynamic interconnected framework that need to be outlined to identify the real innovative nature of the strategic engineering:

- Technological Analysis combining:

    - Data Analytics ↔ AI ↔ Modelling ↔ Simulation

- Data Processing connecting:

    - Big Data from the field → Extracted Information → Measurement of the Impacts on the field of our Decisions → Machine Learning to retune the Models

- Involvement of Decision Makers:

    - Needs and Degrees of Freedom → Strategic Engineering Solution → Use of the Recommendations → Correction of the Objectives

In fact, the first element is strongly dealing with the scientific and engineering world, while the second one gets benefits of the new digitalisation of the world and society. However, it is crucial to understand also the third element, based on involvement of decision makers that is the fundamental component to successfully develop a strategic engineering solution and it strongly relies on the accreditation of the solution. This means that is fundamental to fill up the gap between leaders and analysts, in this case by using strategic engineers with a transdisciplinary background, in order to be able to interact with decision makers and to develop mutual trustiness and common understanding of the complex system to be addressed (Mazal & Bruzzone, 2019).

It is possible to consider another example: the case of a crisis dealing with diffusion of a toxic agent over a populated area due to an industrial accident and let's take as hypothesis that military forces are called to support Civil Protection to evacuate the people, cordoning the dangerous area and conducting the situation assessment. In this case, we can adopt the hypothesis that specific models were developed for these kinds of risks along previous years. In addition is reasonable to expect that there are in place a sensor network, assets and autonomous systems for data collection, as well as capability to acquire data from socials about population feedback. Probably, in this case the Commander has already a valid background on how to manage this type of emergency based on a quite rigid military doctrine for similar scenarios. Therefore, in this case, the real problem could result quite different with quantities of civilians to evacuate that largely overpass available capability because they are resulting from old and too much precautionary models (Bruzzone et al., 1996). In addition, the use of new robotic assets could be really useful to properly assess the situation very quickly, but it could be possible that their use is not really properly addressed in the policies and procedures respect new specific operational capabilities. Based on these hypotheses, a new Strategic Engineering approach could be very useful, but it will be put in place just if the Commanders will develop a full trustiness on this system considering he is aware that his decisions will affect human lives, impact on public opinion, affect political leadership approval of his own actions, his career as well as potential legal threats (Di Bella, 2015).

To use this innovative approach, the Commander should know the reliability of the model, being aware of the consistency of the predictions considering the potential and limitations of this approach within different ranges of operations. To achieve this result, it is fundamental to provide him with strong support to refine and correct plan; this action should be carried out by Strategic Engineers able to interact quickly with him knowing his own language and background; these people should become part of his staff and help him to react on scenario dynamic evolution interacting with the whole team and with the technological decision support architecture.

So, it is evident that the strategic engineering is drastically new approach that combines modern technologies and practical decision making to improve success rate in challenging environments.

6.3 A CRUCIAL PERIOD: COVID-19

In this last year, humanity is literally facing a major modern crisis: COVID-19 pandemic (Meijer and Webster, 2020).

Population in Western countries consider COVID-19 as something completely new as well as the first global disruptive event since the WWII and it should outlined that, nowadays, WWII is a tragedy lost in old memories considering that the large majority of people just born after this war.

Looking at this crisis, it emerges that the western public opinion discovered the possibility of a serious epidemic as a real threat just with COVID-19, even if, just along past 20 years, we experienced SARS outbreak and swine flu pandemic (Heymann, 2004; Versluis et al., 2019). We should add that Spanish flue belongs to contemporary times with quite good records and data, even if it dates back a century ago (Barro et al., 2020): it affected one-third of world population and caused several million casualties; following this terrible event, several other epidemics affected the world. In addition, the risk of biological warfare are known so well since centuries, that Geneva Protocol prohibit use of biological weapons (BW) since 1925, around half million people in China are supposed to be victim of BW due to Japan attacks that by the way backfired even on their own troops (Wilson and Daniel, 2019). From this point of view, almost all major countries developed procedures and plans to face BW threats especially during the Cold War (Christopher et al., 1997). Finally, the public opinion was introduced to pestilence and pandemics risks also by successful Hollywood movies over the years along the years (e.g. The Andromeda Strain, 1971; The Cassandra Crossing, 1976; Outbreak 1995; Contagion, 2011).

Now, based on the real experience of millions of deaths, precise plans for governments, dramatisation by fictions, recent pandemics centred into far East, you should expect that the people and institutions should be able to react effectively to a pandemic, therefore looking to the reactions along this year, with few exceptions, we should realise that most of the countries, including the most advanced and rich ones, were unable to develop and implement an effective strategy.

The detailed analysis of the cause of this failure in terms of strategic capability is out of the scope of this chapter but it is interesting to address few points related to why such failures happen respect something that we know by experience: something that was happening almost two times along past 20 years, some threat that we investigated as mass destruction weapon along Cold World and something that killed over 20 millions of people (someone say 100) just 100 years ago, without mentioning historical threats such as multiple literature books addressing black plague, plague of Justinian and plague of Athens.

To address this point, we should start to outline that there is no way to be ready to face a serious epidemics due to the size of the problem that spreads very quickly among literally millions of people: the resources required to face such a problem, the development and implementation of medical treatments on large scale, as well as the decisions to block the diffusion are overpassing the human capabilities.

However, even if we cannot expect to face these kinds of crises without getting serious damages, it is still evident the lack of strategy in preparing the society to mitigate the impact of current epidemic. In practice, it is possible to define some of the possible sources of this shortfall:

- Short horizon: modern decision makers are looking to very short time horizon; they are really much focused on their objectives (e.g. re-election within 4–5 years, next quarter profits, goal achievement for the current year) to care of problems that are rare but devastating.

- Individual objectives: most of the decision makers are focused on their own goals. So, even during a major crisis they resulted dealing mostly in their personal goals including increasing consensus, discredit opponents, protect personal/regional/national interests, all without being able to identify common global goals, consistent with their own; in fact, they demonstrated often to be unable to identify properly the goals to be achieved to succeed, or they were not confident to be able to face the SARS-CoV-2 virus and preferred to give up about the possibility to block it renouncing to the opportunity to achieve huge consensus and profits.

- Lack of direct experience: despite the recent epidemics, up to 2020, Western countries have pretty limited direct experience on these crises and usually this knowledge resulted in restricted subject matter experts. This created some kind of refusal to accept pandemics as a real threat and as a problem that need resources for being mitigated.

- Confusion: the contingency created confusion among decision makers that demonstrated often limited capacity to finalise decisions and maintain them with persistence, up to verification of their effectiveness; many times this resulted in changing without logic the plan and moving back and forward among different alternatives without being able to finalise them and/or to measure their impact.

- Missing the global picture: in a pandemic there are many interactions among many factors (e.g. vaccine effectiveness and duration vs. treatments, diffusion within schools and infections within public areas), areas (e.g. infected and safe regions, different nations, regions and towns), layers (e.g. health care, economics, population behaviour, etc.). The decision makers addressed usually minor detailed aspects without consideration to the whole picture, resulting often in adopting contrasting measures and changing them ineffectively, as soon as any inconsistency emerges.

In fact, as one of the major examples it is represented by the fact that during COVID-19 crisis, most of the good existing plans for facing pandemics were simply not put in place; for instance, the plan on STRATCOM (Strategic Communications) was often overpassed by multiple declarations of different people with different contrasting theories, often resulting in a major confusion for population and a risky lack of trustiness with respect to institutions and initiatives (Gollust et al., 2020).

We should add that during such a crisis, it is normal to make mistakes: this is a common situation that unfortunately should be consider physiological in a pandemic, so we should be quite tolerant with errors even if they create tragedies. Therefore, despite this kind of tolerance, we should also be firm on the point raised centuries ago by Aurelius Augustinus Hipponensis, stating that "Errare humanum est, perseverare diabolicum" (making mistakes is human, but persevering in errors is diabolical). Indeed, the major problem realised in the management of COVID-19 is related to the fact that the errors committed were not used to correct behaviours, but they were reiterated multiple times and among multiple places, without benefits from lesson learned by countries subjected early to this epidemic.

Based on these considerations, it is evident that even most advanced countries usually don't have proper preparation to face this kind of crucial periods and that during the contingency, the decisions result often inconsistent and ineffective, without correcting wrong decisions and repeating evident errors carried out previously by other decision makers affected early by the virus spreading.

Let's finish on this quick overview by a personal consideration. Currently, the author disagree the current popular idea that nations characterised by lack of democracy could react more effectively and quickly to pandemics and that they are more ready to face crucial periods. Obviously this is a personal point of view, therefore by analysing very crazy decisions taken within few days or weeks by most democratic countries, it results evident that as we take ridiculous decisions quickly, we should be able to take also smart effective decisions at the same speed. For instance, it is not reasonable that we are able to block the direct flights from a contaminated country within 24 hours but, at the same time, we forgot to close borders with other infected areas for months. Obviously, an institution based on a strongman could decide in hours, while the same decision could take few days if it relies on a board, therefore the lack in strategy demonstrated over the COVID-19 crisis is much larger than loosing few weeks and some wrong decisions were implemented and maintained over months (Ruiu, 2020).

6.4 STRATEGIC ENGINEERING AND THE CRUCIAL PERIOD OF PANDEMICS

We already pointed out that there is no magic wand and even strategic engineering cannot solve the problems related to pandemics, such as COVID-19, in one shot. After an outbreak happens in large countries and continental areas, it is mostly impossible to stop diffusion without to put in place drastic measures that are not sustainable for large scenarios and/ or long time; however, it is possible and necessary to develop new strategies to mitigate the crisis evolution and to manage the situation (Espinoza et al., 2020).

Therefore, strategic engineering could address several of these issues and reduce shortfalls, so it results pretty useful in this context due to its own characteristics with respect to specific characteristics of the pandemics corresponding to a real complex system (Bruzzone, 2020).

It is possible to outline that one major lack in the scientific approach to COVID-19 crisis was related to due overfitting analysis on data collected about the epidemics (Bruzzone & Sinelshchikov, 2020); now, it is evident that available data are affected by national security issues, but even consulting teams for decision makers, having access to classified data, should be aware that the reliability of this information is supposed to be affected by quite large confidence bands due to the size of the problem, the not homogenous classification system, the contingency during collection in healthcare infrastructures, the nature of the disease, etc. In fact, we had limitation often even in understanding very basic data, such as the number of casualties. However, the problem in such data analysis was related not only in not properly evaluating its stochastic nature and the experimental errors but especially in willing to use traditional statistics to support decisions by overfitting previous behaviours.

In general, statistics is fine to support decisions, but obviously the phenomena have their own inertia, so if we use statistics to decide if it is better to close or keep open schools, we should accept that we will not extract real suggestions because the statistics rely just on past data, not on the effects on future behaviour of our decisions that are not yet in place.

What we can do for addressing this open issue, it is to take an action and measure consequences to be extrapolated by statistics in next weeks; for instance, in reference to COVID-19, considering elements such as the time to be able to develop symptoms or further infections is supposed to be based on ~10 days, the very high number of asymptomatic subjects (overpassing 50%, plus another 20% of very superficial symptoms), the strong influence of stochastic factors, it results probably that 2–3 weeks minimum are required to a proper evaluation of the impact of our decisions.

This means that a decision that we take today, probably will have some statistical relevance on its positive or negative influence just after several weeks, when probably its effect is already devastating. In fact, it may be possible even to carry out comparative analysis over different places, using data from areas where the virus diffused early, getting benefits of living into a zone that is behind on the pandemic timeline; therefore, also by this approach, the specific nature of the context and the reliability of data could highly affect the quality of the analysis.

To use AI and more sophisticated data analytics respect to statistical analysis could provide significant benefits in terms of correlating different phenomena, identifying and correcting inconsistencies on data, as well as enlarging greatly the situation awareness and scenario assessment by Data Fusion. Therefore, these approaches still relaying on historical data and aren't influenced by the new decisions taken today that will change the scenario evolution over next months.

From this point of view, strategic engineering, by combining AI and data analytics to fuse data and extract more correct information on what happen and what is ongoing, guarantees quite major advantage by adding simulation, able to predict, based on a priori analysis, what will happen respect alternative decisions. Obviously, these simulation models are based on hypotheses and parameters that could be wrong or not properly tuned, therefore the comparison of real evolution with expected results could allow to refine the models by machine learning procedures, getting soon a very reliable decision support system (Bruzzone et al., 2020b).

Indeed, another crucial aspect is the use of the real achievements from the field as support to refine the models, instead than continuously changing parameters to over fit an evolution curve based on basic mathematical models that in reality is continuously modified by the changes on the system caused by our own decisions and other player actions (e.g. we buy vaccine, our allies taking into consideration the risk of lack of doses, put a block on our order).

Another fundamental element of strategic engineering, to be used in this kind of crisis, is the engagement of decision makers in the loop as part of the solution and not just receiving data, because the characteristics of models, algorithms, AI solutions, at least in terms of control parameters and output variables should be commonly shared in terms of precision and confidence band as well as validity ranges for their use.

Therefore, the real added value of strategic engineering in this context is provided not only by supporting crisis management, but also to prepare the background to face such crucial periods, by providing a clear evidence and understandable demonstration of devastating effects of pandemics to decision makers and people respect available resources. These virtual experimentations, as well as the evaluation of costs, risks and effective reaction plans, it is crucial to suggest proper ways to prepare for such a crisis in advance. In some way, by this approach, it turns possible to provide a virtual feedback about these phenomena that are not already known and, in this way, to solicit authorities and decision makers to act in sustainable way to prevent the crisis (Bruzzone et al., 2013).

This is a key point, because while strategic engineering could be useful to design solutions, support operational planning, provide operation support and operational training, the major benefit achievable turn to be the education in strategy development in crucial periods to maximise our capabilities during a crisis and to educate our decision makers.

6.5 EXAMPLE: VESTIGE AND PANDEMICS

VESTIGE (Virus Epidemics Simulation in Towns & Regions for Infection Governance during Emergencies) is a strategic engineering solution developed by Simulation Team of Genoa and it represents the technological element to be adopted as support to face pandemics (Bruzzone et al., 2020b).

This solution has been demonstrated, during last months, in relation to Smart Government issue, while its predecessor, PONTUS (POpulation behaviour, social Networks, Transportation, infrastructures and industrial Urban Simulation), was already employed in strategic planning as part of a large smart city project (Bruzzone et al., 2019b).

In fact, it should be stated that the simulation engine used by VESTIGE is the extension of previous models developed and experimented by the Senior Partners of Simulation Team over past 20 years with specific attention to pandemics and epidemics (Avalle et al., 1995, 1996, 1999; Bossomaier et al., 2009). Indeed, VESTIGE strongly relies on its ability to simulate human behaviours and model the population as well as interest groups. VESTIGE uses intelligent agents to drive the simulation by reproducing individuals and social networks and characterizing people in terms of gender, age, health, education, social level, ethnics, religion, political preferences and other attributes including psychological modifiers such as fear, stress, fatigue and aggressiveness. The people are not only considered as individuals, but even within their social networks including relatives, colleagues and friends and their specific behaviour related to their nature (e.g. manager, employer, housewife, retired person, student, etc.). By this approach it turns possible to simulate the town behaviour under regular conditions as well as during a crisis. In the past, we used this approach to model towns as well as regions (e.g. swine flu pandemic, cyber biological combined national threats, Haiti earthquake, National Spread Civil Disorders within Afghanistan, Nigeria, Katrina Natural Disasters impact) in public domain research (Bruzzone and Massei, 2017).

Indeed, VESTIGE is open to be interoperable with other systems by using HLA (high level architecture); indeed, through HLA it is possible to link several simulators and real systems (e.g. Command and Control, C2) together. The models used within VESTIGE have been developed for crisis as joint ventures among Universities and Spinoffs:

- PANDORA (Pandemic Dynamic Objects Reactive Agents) was a pandemic model devoted to study spreading dynamics of the influenza A virus subtype H1N1 and it was the result of joint cooperation involving United States, European and Australian R&D Centers (MITIM DIPTEM, Dartmouth College, CRiCS).

- CIPROS (Civil Protection Simulator) was a solution able to generate the crisis evolution within a Web framework as support for training and information distribution including different kind of crisis:

  - Epidemics
  - Major flooding
  - Explosions
  - Hazardous material fallout

Obviously, these elements have been adapted to current scenario and the epidemic characteristics within VESTIGE have been tuned on COVID-19 crisis.

VESTIGE scenario configuration is based on data fusion algorithms and AI processing over the region open data (e.g. census, political results and house prices). These data include also the human factors representing the psychological and social parameters that impact on people behaviours (e.g. average number of children, percentage of married couple, etc.) and their reaction to crisis evolution, containment measures and policies.

VESTIGE allows simulating the crisis and so it is possible to estimate the cost/benefit of different strategies for mitigating the crisis or preventing further problems such as school closures, target antiviral prophylaxis, quarantine rules and other mitigation measures even considering impacts on commerce, industrial activity, economy and society activities.

Indeed, by VESTIGE it is possible to decide about different approaches in terms of Alert Level and to correlate epidemic key performance indicators with respect to decision taken to validate their effectiveness.

It is interesting to outline that the effort required to finalise VESTIGE was mostly on the fine tuning of the different models respect COVID-19 case and to finalise verification and validation (V&S) with respect to available data. These actions were pretty fast, considering that the VESTIGE simulation engine, as well as the already available models, resulted very flexible so they don't required major changes on their structure; SIM4Future and Genoa University as active elements of Simulation Team worked effectively on adaptation of models to make VESTIGE available for current COVID-19 crisis within few weeks and obtained a quite usable solution as proposed in Figure 6.1.

Along last year VESTIGE demonstrated realistic behaviour and supported a critical observation of crucial elements related to the current

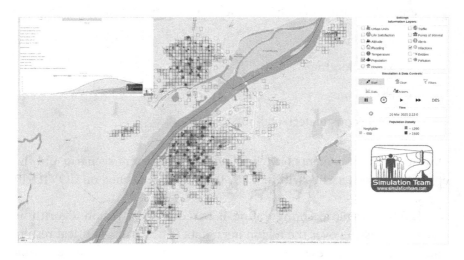

FIGURE 6.1 VESTIGE applied to a region for developing effective strategies against epidemics and impact on population of different policies.

pandemics. Indeed, even if the pandemic evolution remains quite similar over the simulated area, the specific diffusion could be estimated in terms of risk levels and confidence band by considering the stochastic nature of epidemics. Indeed, one of the key elements in the simulation of outbreaks result to be the "patient zero", considering his or her social contacts, work style and free time preferences. These analyses are important to evaluate the impact of social groups to generate hot spots and potential additional risks to be addressed and mitigated. In addition, VESTIGE allows to monitor new infection waves improving the reactivity of the decision makers with help of real data arriving from the field (Bruzzone et al., 2011).

Obviously, VESTIGE confirms that small variation in disease parameters could have large influence on crisis evolution along the entire situation (Roda et al., 2020). Therefore, while this consideration makes quite hard to study the epidemics when many characteristics of the infection are still not well known, it is also possible to conduct risk analysis respect different hypotheses and to identify these parameters based on reverse engineering respect how much the real crisis matches with specific runs. In fact, the simulator allows to apply "what if?" analysis to check different strategies as well as to check different hypothesis on the threat internal parameters.

6.6 EXAMPLE: TOPRO ON HOW TO ADDRESS AN URBAN CRISIS DURING EPIDEMICS

TOPRO (TOwn PROtection) was developed as innovative approach to support education and training of decision makers related to crisis due to CBRNe (Chemical, Biological, Radiological, Nuclear and high-yield Explosive) threats. Indeed, considering the evolution of current COVID-19 crisis, TOPRO was re-focused mostly on biological threats and it adopted a very intuitive and interactive approach for training, education and evaluation of actions adopted by decision makers who deal with the crisis within a town (Bruzzone et al., 2020c).

In reference to strategies development to face an epidemic within a town, it turns to be crucial to define how to prevent and/or manage unexpected events related to population reactions including public events, crowd control, urban disorders and demonstrations; it is evident that in this sense the population reactions to the crisis as well as human factors and economics are fundamental elements as the epidemic nature itself. In addition, it could be necessary also to evaluate effects provided by different technologies (e.g. different testing procedures, autonomous vehicles, different IoT and apps for tracking) so they should be included in the simulation objects.

So, the capability to develop strategies for epidemics relies on being able to create plans and to properly define the skills and capabilities required even respect the population reactions specific of the COVID-19.

Indeed, the actual crisis pushed to adapt the scenario to consider specifically biological contaminations, the related difficulties into handle it and the potential challenges due to urban disorders; in this field it is evident that a proper planning capability, a valid education as well as an effective training are crucial to mitigate the impact of epidemic.

Due to these reasons, it was decided to create TOPRO as solution devoted primarily to education respect pandemics over an urban area to address quickly this crucial problem. The TOPRO simulation engine adopted the innovative paradigm of MS2G (Modelling, interoperable Simulation and Serious Games), so it means that it uses a very intuitive, interactive, immersive virtual environment to make easy to play and to understand the scenario evolution even to players that are not too much expert such as high-level decision makers. In facts, MS2G integrates the classical interoperable high-fidelity simulation with the engagement techniques of serious games for being effective over a large user community.

In the current scenario dealing with protection of a town from the epidemic diffusion, it is necessary to model this complex system and the related people behaviours. So, it was adopted as simulation type a stochastic agent driven simulation able to reproduce people as well as assets operating within and around the town.

TOPRO simulates one week along the diffusion of the infection or of the variant within an area; Simulation Team members were able to finalise quickly TOPRO thanks to the reuse of previous projects and models.

In particular, TOPRO includes following elements:

- A town with its population including different ethnics and cultural backgrounds, living each one within social networks and based on specific behaviour.

- Available resources (police, healthcare, firefighters etc.) and additional armed forces to be assigned to different roles (e.g. checkpoints, decontamination activities, crowd control and health care extra support)

- Specific terrain and boundary conditions including roads, streets, commercial centres, squares, parks, churches, rivers, hills, mountains affecting movements and behaviour of people.

- Presence of an external factor that directly on in combined way, could inject additional challenges and force to take immediate actions for remediation. Indeed, some of these criticalities could be introduced by arrival of refugees or arrival people infected by a new virus variant as well as tentative to force check controls.

- Presence of critical infrastructures in the area that even during lockdown requires to have a safe access and protection; in the current case we introduced a power station and the potential risk of disorders.

- Country border in the area introducing crucial issues such as immigration procedures, smuggling and criminal activities, commuters travelling and border patrols. In this case, people scared about the diffusion of the virus on a side of the border could develop demonstrations and riots, as well as internal people if they feel at risk for some decision affecting their health safety or economic conditions.

People are driven by agents and they move around by different ways, including walking, using bicycles, bikes or cars; obviously this makes even more complex to take care of tracking people and to intercept them in case they are suspected to be infected.

TOPRO proposes these elements as part of a serious game devoted to educate decision makers in containment of a biological crisis and in responding to critical events to protect of a town from epidemic. The user could dispose resources, introduce procedures, activate checkpoints and testing facilities in the town or on the border between the safe are and the infected one. The evaluation and impacts of demonstrations and other actions are simulated automatically by the agents within the TOPRO simulation.

The user has control of his resources and units including:

- Police, trained to deal with civilians, but with limited capability to hold incoming flows of people. These units are relatively fast within the city and deploy quite quickly. Police units use cars and motor bikes, plus helicopters, which grant them high mobility, but just for surveillance. These units are available in town area.

- Military units not trained to deal with civilian, but with higher force to handle people even if with risk to create tensions. In additional there are special units with specific capabilities in terms of decontamination and testing. Military move mostly by trucks at lower speed on the city and they require additional time for deployment into the area due to the fact that they arrive from outside the town, with the exception of few units preliminary deployed in the city.

- Autonomous assets to be used in order to conduct surveillance and search in urban areas and bordering rural zone (e.g. unmanned aerial vehicles)

- Containment forces trained and specialised for riots and demonstrations. Their characteristics are similar to previous ones, there are limited capabilities in the town, while additional ones could be required from outside and require extra time to reach the city.

- Healthcare units devoted to deal with infected people and carrying out test.

TOPRO proposes different use modes to the player:

- Observer: TOPRO shows predefined scenarios and it takes care of moving automatically all the assets. So, in this mode, the user turns to be just a virtual observer able to move around into the world to better observe the actions. In this way, the decision makers see directly the dynamics of the events, understand the specific planning to manage some kinds of problems, while they develop their own scenario awareness based on symptoms and events.

- Free play: This mode is devoted to the familiarisation on the scenario. In this case, the players are enabled to control the units and assets in order to become more familiar with their capabilities, mobility, constraints and delays as well as with other specific aspects of the operations. In this mode the user can test different initial conditions and then to test into the simulation their effectiveness respects the classical scenario of crowd pressure over the checkpoints to enter into the town.

- Competition: In this mode, the players start from a given situation, automatically generated by TOPRO, and they should develop a strategy to contain the threats by using controlled units. TOPRO provides an evaluation of the performance respect the difficulty levels of the scenario based on the improvements achieved by each player respect other ones and/or respect his previous simulations.

- Analysis, which allows evaluating the effectiveness and efficiency of actions conducted by a player respect some infected people penetrating the area and the tracking procedures. In fact, in this mode, it is estimated the time required to identify, track and block these infection agents as well as the percentage of successful containment obtained respect the quantity of involved assets and costs for the entire operation.

As mentioned, the immersive property is important element for efficient education and training on gaming; in this case the game could simply run on smart phones, but it is also possible to integrate it on laptop, on virtual reality headsets as well as on an interactive whiteboard for blended education as in Figure 6.2.

TOPRO has been used for education and training and it is currently subjected to further developments respect new applications and scenarios.

FIGURE 6.2 TOPRO Virtual Interoperable Intuitive Serious Games to educate top decision makers with respect to biological hazards and CBRNe threats.

6.7 EXAMPLE: T-REX FOR BEING READY FOR HYBRID AND CYBER THREATS

T-REX (Threat network simulation for REactive eXperience) is a framework developed to investigate hybrid threats over a region and how to reduce related vulnerabilities (Bruzzone and Massei, 2017). Indeed T-REX addresses the hybrid menaces with respect to authorities and companies managing critical infrastructures such as ports, power plants, and desalination facilities. T-REX is also based on MS2G simulation and it allows to develop exercise to develop strategic engineering approach in engineers and decision makers by interactive experience (Bruzzone and Sinelshchikov, 2020).

Here the proposed scenario is centred within a region including a major city and four towns within a desertic area with critical elements in power, oil and water resources (Bruzzone et al., 2017c). Hybrid threats are including multiple alternatives such as cyber-attacks, media attacks, fake news, conventional attacks, diplomatic actions, financial actions, etc. The use of strategic engineering in this context is fundamental to understand how the different countermeasures could reduce the vulnerabilities as well how to design new polities, new technologies, new organisations and new assets to prevent and mitigate risks.

T-REX proposes the case of an hypothetical large corporation operating over multiple offices distributed in five towns within the desertic region, it consider to have fixed work stations, laptop and adoption of extensive smart work procedures. The corporation controls a port terminal, a refinery, a power plant, a desalination facility and strategic commodities such as water distribution, power grid and a tank farm for oil storage.

It is evident that in this scenario there are multiple vulnerabilities that could be attacked by cyber actions, conventional attacks as well as attacks on media, for instance promoting idea that water is contaminated or energy resources are compromised, even if they are still safe; so also in this case, the population behaviour and its reactions result to be a key element of the scenario.

The decision makers could act along multiple degrees of freedom to modify the situation and to verify the effectiveness of new measures respect vulnerabilities.

The parameters that could be controlled include, among the others: Employer organisation, office structures, corporation network, smart working in terms of frequency and diffusion (e.g. days/week), different levels of presence and preparation of employers in terms of ability to face cyber threats, cyber security units capabilities and number, antivirus diffusion and efficiency, network protection, size and effectiveness of communication division, resilience of water, power, port infrastructures, refinery, grids, etc. Vice versa, the output variables include effective vulnerability measured on power, oil and water resources, survivability and resilience level, quantity of experienced major vulnerabilities, quantity of experienced minor vulnerabilities, how much time is required to break down the services, mean time to restore the service, costs of the attacks, costs of defensive actions, quality of service, impact on reputation, cyber asset level in terms of integrity/availability/confidentiality, probability to prevent the crisis, probability to identify and track the attackers.

Boundary conditions includes time and duration of attacks, environmental conditions, costs for the attackers, attacker capabilities, service levels required by users of services (e.g. water/power/oil), percentage of wrong behaviours by employers of the corporation, percentage of spills of information, virus injection mode, virus resilience, virus infectivity, undefended IoT, antivirus diffusion outside the corporation, strong/weak antivirus diffusion out of the corporation etc.

The audience is required to address the degrees of freedom for strategic managers as well as to define priorities on target functions and to test them within the T-REX virtual simulation environment. The results achieved in this way are obtained during the exercise by interactive discussions in the team and with expert as well as through review of the scenario simulations conducted on T-REX.

In the Figure 6.3 is proposed the architecture and the exercise scheme including DCS (digital control systems) models of the plants and corporation network (ICT).

T-REX uses intelligent agents (IA) to interest groups corresponding to authorities and corporation division as well as the population of each town. In addition, IA controls the offensive/defence units and other operational units devoted to carry out specific actions. Also in this case, the Simulation is based on MS2G paradigm (Modelling, interoperable Simulation and Serious Games) and it is based on agent driven discrete event stochastic simulation.

In the Figure 6.4, it is proposed the simulator proposal to the player using the SPIDER (Simulation Practical Immersive Dynamic Environment for Reengineering) framework; however, the systems use Simulation Team engine that is scalable from PC to CAVE (Cave Automatic Virtual Environment).

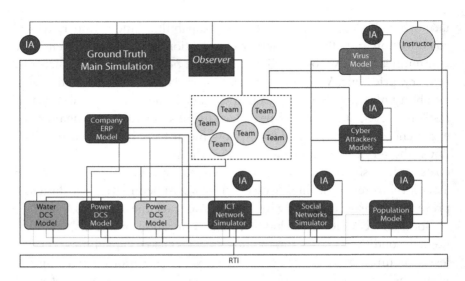

FIGURE 6.3   General architecture for the use of interoperable simulation in hybrid threats.

This exercise is devoted to introduce decision makers and managers to strategic engineering and develop their skills by interactive virtual experiences. Indeed, each team needs to address the different aspects of the scenario and to develop a clear understanding of the whole situation and different threats. Obviously, a major goal is the capability identify the vulnerabilities and problems and to apply strategic management to this specific case study.

Along the exercise, the team is required to move forward in problem analysis and to identify the most critical set of parameters that could

FIGURE 6.4   T-REX simulation including virtual representation of physical world and cyber space.

reduce vulnerabilities in effective way and respect the risks, constraints of costs, resources. In fact, the simulator generates the scenario dynamics and related data generated, obviously considering decisions of the team. The team needs to identify what and how to collect data and to define how to conduct the analysis to identify inconsistencies and red spots that suggest presence of emerging criticalities and/or threats. Obviously the use of machine learning, AI and data analytics is a strategic advantage in the identification of correlations useful to support decisions. In addition, each team needs to correct its own decisions based on scenario evolution and to finalise decisions on prevention and vulnerability reductions as well as re-planning due to contingencies. T-REX manage the scenario eventually by injecting additional challenges, but also providing information spills about the threats as well as disclosing symptoms of new incoming problems.

Debriefing and After Action Review (AAR) on the exercise are carried out with the class over the T-REX framework to identify lesson learned. Each team is evaluated in terms of his quick reaction capability and the ability of the members to use strategic engineering methods to identify as early as possible the criticalities and potential threats within the scenario. Success of the strategic management approach is based on the achievements on target functions and on KPIs (key performance indicators) as well as in terms of capability of the team to develop a comprehensive view of the scenario.

6.8 EXAMPLE: ALACRES2 VIRTUAL LAB TO IMPROVE PORTS IN TERMS OF SAFETY AND SECURITY

ALACRES2 is a virtual lab created in Italy and France to address crucial issues related to port safety and security (Bruzzone et al., 2019c). Indeed, ALACRES2 should act as permanent laboratory capable of identifying, testing and validating integrated new procedures to prevent and mitigate major accidents or claims occurring during the loading and unloading phases of goods in the port with special attention to the combined effects of events evolving concurrently (e.g. ferry boat operations, container operations, dangerous material handling).

In this case, creating an innovative virtual lab to put together the decision makers with strategic engineers aims to identify univocal management protocols, organisation structures and proper behaviours to assist the improvement of ports and marine logistics. In fact, if we look at the existing situation in ports, it is clear that accidents often occur due to the complexity of the context. The recent case of accident in Tianjin (2015) and

Beirut (2020) corresponds to two of the largest non-war explosions along human history:

- Tianjin explosion, August 12, 2015, 800 tons of ammonium nitrate, explosion corresponding to 336 tons of TNT, 173 casualties, 2 km destruction range, 9 billion USD damages

- Beirut explosion, August 4, 2020, 2,750 tons of ammonium nitrate, explosion corresponding to 1.2 kilotons of TNT, 207 casualties, 300,000 homeless after explosion, 15 billion USD damages

Now these specific two events (in past an accident was also in port of Texas City, April 16, 1947, 2,300 tons of ammonium nitrate, 581 casualties, largest not nuclear explosion in United States) are result of very improper storage characteristics that in developed countries are supposed to be not possible nowadays, therefore ports are still very critical respect accidents and future developments in terms of logistics flows, urbanisation, demographics could further reinforce the necessity to address these crucial elements represent safety and security.

In facts, mentioned events are symptomatic of the complexity due to the multiplicity of active subjects, in high operational density scenarios, within ports managing enormous flows at high speed and that are strictly surrounded by urban areas that have grown over the years around the port.

This is another example where the strategic engineering allows developing experience and understanding complex system respect crucial conditions as well as potential future risks that are rare, but characterised by huge impact. Indeed the accident in ports cannot be tested on the field, if not in very superficial ways. This confirms the need to set up a virtual laboratory that uses simulation techniques and develop strategies for decision makers.

One need is certainly that the ALACRES2 simulation allows recreating the emergency scenario as a whole, including its immersive and intuitive representation in virtual reality, allowing the operators involved in the experimentation/training to operate and simulate their own action. From this point of view, the realism, including human factors, visual and sound, the climatic and surrounding conditions are often very critical in the generation and evolution of emergencies.

A clear need for the development of this laboratory is the involvement of both expert simulation partners who will develop the virtual reality

FIGURE 6.5 ALACRES XR solutions overlapping the simulation results to a physical nautical map of the port to share it with decision makers.

environments and fine-tune the laboratory tools, and operational partners, who share their experience in the field, in ports and at sea together with the relative emergency procedures. This activity is a clear need not only for the development of the ALACRES2 virtual laboratory, but also for the test and use envisaged in the project.

In effect, ALACRES2 uses XR (eXtended Reality) and simulation to immerse the decision makers in the problem and show the effects of vulnerability on safety or security as well as the advantages and disadvantages of alternatives procedures (see Figure 6.5). ALACRES2 reproduces not only assets and operations, but include models of the different operational figures called to carry out activities in ports (e.g. custom, passengers, riggers, crane operators, managers, etc.) as well as manage emergencies (e.g. firefighters, coast guard, health care).

ALACRES2 allows to test new behavioural protocols, new operating standards, new emergency monitoring and technological solutions, as well as control procedures. In particular, the ALACRES2 laboratory aims to analyse the behavioural procedures and protocols of the most critical subjects, namely:

- Leaders of the chain of command and/or of the operational management centres, or those who are appointed to manage a long-lasting emergency condition (e.g. widespread and prolonged fire, uncontrolled spillage in water, evolving toxic cloud, etc.);

- Operational subjects in charge of first aid activities aimed at stemming the emergency and/or reducing the causes that generated the indicator (e.g. firefighters, emergency operators, etc.).

For the different types of logistics flows, ALACRES2 simulates the evolution of the scenario until the injection the cause of a potential emergency in order to check capability to prevent the event and/or mitigate its impact.

In this sense, ALACRES2 allows you to test the conditions that generate the accident and which normally have to do with human responses and behaviours in conditions of stress, work overload, redundancy or absence of information, etc.

6.9 EXAMPLE: ITWETS AND WATER STRATEGIES

Nowadays we are experiencing COVID-19 pandemic, therefore it is evident that our contemporary health challenges are strongly related to water availability as cornerstone for health, hygiene and sanitation. This aspect is probably drastically underestimated in developed countries, but it is very crucial for developing areas where fresh water is scarce. Prof. Alessandro Leto, director of Water Academy, said that the major weapons against this pandemic as well as other diseases are water and soap (by the way production of soap is based on water use and it is another challenge for developing areas) because hygiene strongly relies on them. In addition, it is evident that fresh water is the basic resource for human life, not only for drinking, but even to sustain food production. Water is also a fundamental element for almost all industrial activities including mining, production power generation. Unfortunately, fresh water is a scarce resource on the planet, while pollution and climate change are currently affecting this strategic resource, so we are facing a major risk of a terrific crisis dealing with water, potentially hitting many of the consolidated systems on which our institutions are built, both at the international and national level.

Indeed, the water crisis is partially already here and has deep and increasing impact on geopolitical balance; this situation could lead to big tensions and even wars, and it will be necessary to develop effective water strategies and activate new policies and huge investments. In facts it is evident that the human progress strongly relies on access to water to the entire world population; therefore, while our planet is covered for 70% by water, only 3% of this resource is sweet water, included the vast quantity present in the "eternal ice" (Arctic and Antarctic). In practices, the earth is short of surface sweet waters and, in addition, it is not equally distributed in space, nor in time. As result of these conditions, we have to face the double challenges of a Geometric Demographic Growth (Malthus, 1798) and the growing need of water for agricultural products increasing risks of food security crises in many territories. This means that in future we could face a new crucial scenario where we had to decide between using water for hygiene and health, or for irrigating the crops for feeding population (Leto, 2014).

Due to these reasons, we have developed ItWets as simulation framework to prepare decision makers to address issues related to strategies for water (Bruzzone, Leto and Scotto, 2020d). ItWets proposes a serious game based on discrete event stochastic simulation based on MS2G paradigm. The scenario is related to develop a large dam dealing with two countries that have extreme need of water and are sharing a scarce source while multiple problems could arise in design, construction, activation, diplomatic level, presence of insurgents, diffusion of diseases, social tension, security issues, etc. The case is inspired by the Grand Ethiopian Renaissance Dam (Mulat and Moges, 2014). In ItWets, each player has to take crucial decisions to face contingencies as well as to manage this large program for international sponsors and in cooperation with regional authorities.

The Serious Game is part of an educational path developed jointly by Simulation Team, SIM4Future, STRATEGOS at Genoa University and Water Academy and it was already proposed in several international contexts.

The final goal is to prepare the new decision managers for water resources and to introduce them into the complexity of this framework; in addition, we are focusing on teaching how strategic engineering could be very important in addressing crucial challenges for our future. The advantage to create such kind of Serious Game including many factors and models, but very intuitive and user friendly, running on a smartphone and providing to players a self-assessment on a major water crisis, results to be very interesting. This makes it possible also to support diffusion of the cultural background on water strategies within top-level decision makers and to educate public opinion about how important is this topic for the future of humanity.

6.10 SUMMARISING

This overview on strategic engineering respect crucial periods and the example proposes confirm the potential of this approach. In facts, the modern Simulation, AI and Data Analytics combined all together and supported by new transdisciplinary teams could obtain an effective engagement of Strategic Decision Makers and improvement on their processes. This synergy and cultural background evolution is a great opportunity to use new methodologies enabled by technology advances to face critical situations based on real and dynamic Data.

So, from this point of view, the strategic engineering seems to be a major step forward for developing new capabilities to tackle major future challenges and prepare new generations of leaders.

REFERENCES

Amico, V., Bruzzone, A. G., & Guha, R. (2000, July). Critical issues in simulation. In SUMMER COMPUTER SIMULATION CONFERENCE (pp. 893–898). Society for Computer Simulation International, 1998.

Avalle L., Bruzzone A. G., Copello F., Guerci A. (1996). "Determination and Quantification of Functional Parameters Relative to Contamination Vector Logic in an Epidemic Simulation", Proceedings of ESS96, Genoa, October 24–26, Italy.

Avalle L, A. G. Bruzzone, F. Copello, A. Guerci, P. Bartoletti (1999). "Epidemic Diffusion Simulation Relative to Movements of a Population that Acts on the Territory: Bio-Dynamic Comments and Evaluations", Proceedings of WMC99, San Francisco, January.

Avalle L., Bruzzone A. G., Copello F., Guerci A., & A. Scavotti (1995). "Preliminary Analysis for the Creation of a Territorial Epidemic Simulation Model", Proc. of ESM95, Praha, June 5–7.

Barro, R. J., Ursúa, J. F., & Weng, J. (2020). The coronavirus and the great influenza pandemic: Lessons from the "Spanish flu" for the coronavirus's potential effects on mortality and economic activity (No. w26866). National Bureau of Economic Research.

Bossomaier, T., Bruzzone, A. G., Massei, M., Newth, D., & Rosen, J. (2009). "Pandemic Dynamic Objects and Reactive Agents". Proceedings of International Mediterranean Modeling Multiconference, Tenerife, Spain, September.

Bruzzone, A.G. (2018a). "Strategic Engineering: How Simulation could Educate and Train the Strategists of Third Millennium", Proceedings of CAX Forum, Sofia, September.

Bruzzone, A.G. (2018b). "MS2G as Pillar for developing Strategic Engineering as a New Discipline for Complex Problem Solving", Proceedings of I3M and Keynote, Budapest, September.

Bruzzone, A.G. (2020). "Experiencing Strategic Decision Making", Proceedings of SummerSim.

Bruzzone, A. G., Di Matteo, R., & Sinelshchikov, K. (2018). Strategic Engineering & Innovative Modeling Paradigms. In Workshop on Applied Modelling & Simulation (p. 14).

Bruzzone, A. G., Leto, A., Scotto P. (2020c). "Modeling, Interoperable Simulation & Serious Games to Educate how to develop Water Strategies during a Crisis", Proceedings of NATO CAX Forum.

Bruzzone, A.G., & Massei, M. (2017). Simulation-based military training. In Guide to Simulation-Based Disciplines (pp. 315–361). Springer, Cham.

Bruzzone, A. G., Massei, M., Fabbrini, G., Gotelli, M., Bella, P.D., & Pusillo, L. (2019a, July). Libra ad bellum novum: a political and military escalation in the near east as scenario for support advanced strategic decision making. In Proceedings of the 2019 Summer Simulation Conference (pp. 1–9).

Bruzzone, A. G., Massei, M., Madeo, F., Tarone, F., & Petuhova, J. (2011). Intelligent Agents for Pandemic Modeling. In Proceedings of the 2011 Emerging M&S Applications in Industry and Academia Symposium, SCS, April, pp. 23–30.

Bruzzone, A. G., Massei, M., Poggi, S., Dallorto, M., Franzinetti, G., & Barbarino, A. (2013). "Quantitative Simulation of Comprehensive Sustainability Models as Game Based Experience for Education in Decision Making", Proceedings of I3M2013, Athens, Greece.

Bruzzone, A. G., Massei, M., & Sinelshchikov, K. (2019b, July). Application of blockchain in interoperable simulation for strategic decision making. In Proceedings of the 2019 Summer Simulation Conference (pp. 1–10).

Bruzzone, A. G., Massei, M., Sinelshchikov, K., Fadda, P., Fancello, G., Fabbrini, G., & Gotelli, M. (2019c). Extended reality, intelligent agents and simulation to improve efficiency, safety and security in harbors and port plants. In 21st International Conference on Harbor, Maritime and Multimodal Logistics Modeling and Simulation, HMS 2019 (pp. 88–91).

Bruzzone, A.G., & Signorile, R. (1998). Simulation and genetic algorithms for ship planning and shipyard layout. Simulation, 71(2), 74–83.

Bruzzone, A. G., & Sinelshchikov, K. (2020). Strategic Management and Simulation: a Live Interactive Experience for experiencing the Methodology, proceedings of SpringSim.

Bruzzone, A. G., Sinelshchikov K., Gotelli M., Fabbrini G. (2020b) "Town Protection Simulation", Proceedings of MAS2020, September.

Bruzzone, A. G., Sinelshchikov K., Massei M. (2020a) "Epidemic Simulation based on Intelligent Agents", Proceedings of I_WISH, September.

Christopher, L. G. W., Cieslak, L. T. J., Pavlin, J. A., & Eitzen, E. M. (1997). Biological warfare: a historical perspective. JAMA, 278(5), 412–417.

Clymer, A. B. (1993, December). Applications of discrete and combined modeling to global simulation. In Proceedings of the 25th conference on winter simulation (pp. 1135–1137).

Di Bella, P. (2015). "Present and Future Scenarios and Challenges for M&S in terms of Human Behavior Modeling", Proceedings of I3M & Keynote Speech, Bergeggi, September.

Eisenman, J., & Kurlantzick, J. (2006). China's Africa strategy. Current history, 105(691), 219–224.

Espinoza, B., Castillo-Chavez, C., & Perrings, C. (2020). Mobility restrictions for the control of epidemics: When do they work?. PLoS One, 15(7), e0235731.

Ghahramani, Z. (2015). Probabilistic machine learning and artificial intelligence. Nature, 521(7553), 452–459.

Ghani, N. A., Hamid, S., Hashem, I. A. T., & Ahmed, E. (2019). Social media big data analytics: A survey. Computers in Human Behavior, 101, 417–428.

Gollust, S. E., Nagler, R. H., & Fowler, E. F. (2020). The emergence of COVID-19 in the US: a public health and political communication crisis. Journal of Health Politics, Policy and Law, 45(6), 967–981.

Heracleous, L., & Werres, K. (2016). On the road to disaster: Strategic misalignments and corporate failure. Long Range Planning, 49(4), 491–506.

Heymann, D. L. (2004). The international response to the outbreak of SARS in 2003. Philosophical Transactions of the Royal Society of London. Series B: Biological Sciences, 359(1447), 1127–1129.

Isaksson, A. J., Harjunkoski, I., & Sand, G. (2018). The impact of digitalization on the future of control and operations. Computers & Chemical Engineering, 114, 122–129.

Kitchin, R., Coletta, C., Evans, L., Heaphy, L., Perng, S. Y., et al. (2016). How vulnerable are smart cities to cyberattack.

Lehmann, A., & Wang, Z. (2019). Verification, Validation, and Accreditation (VV&A)—Requirements, Standards, and Trends. In Model Engineering for Simulation (pp. 101–121). Academic Press.

Leto, A. (2014). The role of sustainable and responsible development to preserve and increase the 3S: sovereignty, safety and sustainability, Water Academy Report.

Li, B. H., Zhang, L., Li, T., Lin, T. Y., & Cui, J. (2017). Simulation-based cyber-physical systems and Internet-of-Things. In Guide to Simulation-Based Disciplines (pp. 103–126). Springer, Cham.

Malthus, T. (1798). An Essay on the Principle of Population. An Essay on the Principle of Population, as it Affects the Future Improvement of Society with Remarks on the Speculations of Mr. Godwin, M. Condorcet, and Other Writers, 1798, Cosimo.

Mazal, J., & Bruzzone, A. G. (2019). NATO needs of Future Strategic Engineers. In Workshop on Applied Modelling & Simulation 35.

McLeod, J. (1972). Simulation today—and yesterday. Simulation, 18(5), 1–4.

McLeod, J., & McLeod, S. (1995) Mission Earth and the Big Bird from the Ashes. Simulation, SCS, 64(1), June 1.

Meijer, A., & Webster, C. W. R. (2020). The COVID-19-crisis and the information polity: An overview of responses and discussions in twenty-one countries from six continents. Information Polity, (Preprint), 1–32.

Mulat, A. G., & Moges, S. A. (2014). Assessment of the impact of the Grand Ethiopian Renaissance Dam on the performance of the High Aswan Dam. Journal of Water Resource and Protection, 2014.

Ozaki A., Furulchi M., Nakajima K., Tanaka H., & Abe K. (1996). Parallel Car traffic Simulation Based on Space-Time Object Genoa, October.

Pal, N. R., Corchado, E. S., Kóczy, L. T., & Kreinovich, V. (2015). Advances in Intelligent Systems and Computing.

Piera, M. À., Narciso, M., Guasch, A., & Riera, D. (2004). Optimization of logistic and manufacturing systems through simulation: A colored Petri net-based methodology. Simulation, 80(3), 121–129.

Roda, W. C., Varughese, M. B., Han, D., & Li, M. Y. (2020). Why is it difficult to accurately predict the COVID-19 epidemic? Infectious Disease Modelling.

Ruiu, M. L. (2020). Mismanagement of COVID-19: Lessons learned from Italy. Journal of Risk Research, 23(7–8), 1007–1020.

Turing, A. M. (1948). Intelligent machinery.

Turner, C. J., Hutabarat, W., Oyekan, J., & Tiwari, A. (2016). Discrete event simulation and virtual reality use in industry: New opportunities and future trends. IEEE Transactions on Human-Machine Systems, 46(6), 882–894.

Versluis, E., van Asselt, M., & Kim, J. (2019). The multilevel regulation of complex policy problems: Uncertainty and the swine flu pandemic. European Policy Analysis, 5(1), 80–98.

Williams, E. J. (1998). Verification and validation in industrial simulation. In Summer Computer Simulation Conference (pp. 57–62). Society for Computer Simulation.

Wilson, J. M., & Daniel, M. (2019). Historical reconstruction of the community response, and related epidemiology, of a suspected biological weapon attack in Ningbo, China (1940). Intelligence and National Security, 34(2), 278–288.

Wu, Y. J., & Wang, Y. (2008). Google-Map-based online platform for arterial traffic information and analysis. In 87th Transportation Research Board Annual Meeting, Washington, DC.

CHAPTER 7

Cyber Security

Evolving Threats in an Ever-Changing World

Roberto Dillon[1], Paul Lothian[2], Simran Grewal[3] and Daryl Pereira[4]

[1]School of Science and Technology,
James Cook University Singapore, Singapore
[2]KPMG Singapore, Singapore
[3]KPMG Singapore, Singapore
[4]KPMG Singapore, Singapore

CONTENTS

DOI: 10.1201/9781003148715-7

7.1 INTRODUCTION

The astonishingly fast rate at which technology is evolving has several implications. Not only new tools and opportunities to improve workflows and overall daily life are constantly emerging and becoming widespread but, likewise, a darker side of progress is also evolving in parallel at the same speed, if not even faster. In fact, as technology becomes more powerful and, at the same time, easier to use, so becomes the expertise of cyber criminals. Most importantly, new technologies designed to perform a complex set of tasks, often depending on other tools or services, do potentially offer a multitude of attack vectors that can be exploited to disrupt, steal or gain unauthorised access to data, functionalities or services. Unfortunately, it is safe to say that with new technologies come new, hidden vulnerabilities.

As discussed in detail in Section 7.2, the scenario we were living in before the COVID-19 pandemic was already critical. Accenture, in its 2019 annual report, pointed out how threat actors were growing in sophistication and professionalism, paying an increasingly close attention to global event with the intent of using these as additional opportunities to exploit and target people in different ways (Accenture Security, 2019, p. 8).

With an internet-active population growing at an unprecedented rate, doubling from 2 billion to 4 billion between 2015 and 2018 (Morgan, 2019a, p. 3), and a steadily increasing number of Internet of Things (IoT) devices connected to the world wide web, which is estimated to reach 50 billion by the end of 2020 (Evans, 2011, p. 3), targets are in no short supply.

From a strictly monetary perspective, cyber-crime actors stole, or provoked damages for, a massive $3.5 trillion since 2001 (Clement, 2020) thanks to a relentless storm of ransomware, stolen identities and botnet-based DDOS attacks across all industries and continents.

With this set-up at the beginning of 2020, it would have been natural to expect cyber threats to become more challenging even in an otherwise safe and calm world. Unfortunately, the explosion of COVID-19 was an unexpected natural disaster during our lifetime and contributed to affecting the cyber landscape in many indirect and unpredictable ways, painfully pointing out old and new weaknesses alike in our growing technologically dependent society.

7.2 WHERE WE ARE COMING FROM

It is undeniable that a significant digital transformation was already underway even before the pandemic. Attacks based on phishing and ransomware were a significant threat. Companies were moving to cloud and continued to try to fill the cyber talent gap. Cyber leaders had a voice at the exec table. Sectors such as healthcare, education and manufacturing were increasingly targeted by attackers, in addition to the financial sector.

In 2019, cyber-attacks topped the list of business risks in North America and Europe and came second only to natural catastrophe in East Asia and the Pacific (World Economic Forum, 2019, pp. 12–13, 16–17, 22–23). In 2019, 69% of CEOs said that a strong cyber strategy is critical to building trust with key stakeholders, up from 55 in 2018. Seventy-one percent of CEOs say that their organisation sees information security as a strategic function and a source of competitive advantage. Cyber security remained one of the top five risk in 2019 compared with 2018 (KPMG International, 2019, p. 6).

Nonetheless, even though digital transformation was underway and was a priority for directors, CEOs and senior executives, according to Tabrizi, Lam, Girard, and Irvin (2019), around 70% of digital transformation initiatives did not meet their goals due to efforts being more technology-led rather than business-led. Companies also continued to upgrade and replace their legacy systems.

From a broad perspective in 2019, there were several critical issues to consider: Skills shortage, artificial intelligence (AI), data privacy compliance, fraud risk and cyber risk, authentication and phishing (KPMG US, 2019). Headlines and anniversaries worth remembering in 2019 include the fifth birthday of the NIST Cybersecurity Framework. The Heartbleed vulnerability (CVE-2014-0160) also celebrated its fifth birthday since its public announcement, and yet, as pointed out by Forrester (2019a), many systems still remained unpatched. In its first year, GDPR fines kicked in with some reaching tens of millions of Euros (Small and De Fonseka, 2019). Last but not least, high-profile members of the group deploying "Bugat" malware were indicted (Department of Justice, 2019).

At the same time, as companies leveraged public cloud, they also saw a rise in costs. So they started to evaluate hybrid models to try to reach more predictable costs, according to Continuity Central (2019a). They also started to use AI to predict downtime. As companies progressed on their cloud journey, many moved to multi-cloud and paired up with the large cloud service providers (Dignan, 2019a). AI, analytics and IoT became key value-added services for these providers.

In the end, to properly contextualise this complex scenario and understand the dramatic effect and consequences produced by the pandemic, we need to have a proper understanding of the different attack vectors as well as of the ongoing geopolitical and business shifts.

7.2.1 Phishing and Ransomware

Ransomware hit the headlines back in 2017 with some of the largest global attacks and insurers citing it as the leading cause of claims in 2018 (Ng, 2019). Ransomware took a new turn in 2019 when more than 70 state and local governments were victims of ransomware attacks. Many took the decision to pay the hackers rather than pay to rebuild their systems. Cryptojacking gained ground on ransomware and phishing and bitcoin illegal activity approached size of US/Europe illegal drugs market (Morgan, 2019b).

Cyber-crime continued in 2019 as advanced phishing kits were made available on the dark web. There was an increase in mobile attacks as users went mobile. Home automation started to take off and attackers exploited the Internet of Things. AI was considered a dual-use cyber technology, as it was used by both attackers and defenders (von Gravrock, 2019).

Some of the highest DDOS attacks in terms of packet-per-second were seen in 2019, with some exceeding 500 million pps (Shani, 2019).

There were also cloud attacks on financial services companies, ransomware attacks on manufacturers, IP attacks on the pharmaceutical and automotive sectors, DDoS attacks on grid operators, and campaigns against European industrial firms (Osborne, 2019; Verizon, 2019, pp. 62–64).

7.2.2 Cyber and Geopolitics

NERC warned that suspected Russian hacking group was snooping on electrical utilities' networks. Also, in 2019 seventeen US utility companies were targeted by a Chinese state-sponsored hacking group (Center for Strategic and International Studies, 2021). The Council on Foreign Relations tracks publicly known state-sponsored incidents against the United States, which is focused on China, Russia, Iran and North Korea. In 2019, there were 76 operations, mostly acts of espionage (Council on Foreign Relations, 2021).

7.2.3 Recovery Was Underfunded and Insurance Grew

Business continuity was typically underfunded and under-resourced, according to Continuity Central (2019b), with 52.4% saying that this will be their top challenge in 2019. Cyber insurance direct written premiums grew by 12% in 2019 to over $2.2 billion versus 8% growth in 2018 (Fitch Ratings, 2020).

7.2.4 Cyber Skills Shortage but Women Make Impact

Sixty-five per cent of organisations reported a shortage of cyber security staff, with a shortage of around 4 million cyber security professionals worldwide [(ISC)^2, 2019a)]. Women in cyber security continued to grow with an (ISC)^2 survey [(ISC)^2, 2019b)] estimating that the percentage of women in cyber security is roughly 24% and that they typically have higher levels of education, and are finding their way to leadership positions in higher numbers, even though there was a pay gap.

Among the organisations whose leadership believes that their peers are investing significantly, 69% of them are treating security awareness as a top priority. However, over 75% of security awareness professionals were part-time (SANS Security Awareness, 2019).

7.2.5 Budgets Increased with Some Large M&A

Almost half, or 47%, of CISOs were determining how to control security spending based on organisational security outcome objectives (Cisco, 2019). The State of Georgia invested in a US$100 million Cyber Centre which was one of the largest single investments in a cyber centre (Georgia Technology Authority, 2019). In 2019, the US budget for cybersecurity-related activities was US$16.9 billion (Office of Management and Budget, 2020, pp. 265–269). Analysts projected that global security spending on hardware, software and services will be US$103 billion (Dignan, 2019b), with United States, China, Japan and the United Kingdom leading the market. Broadcom bought Symantec's enterprise security business for US$10.7 billion (Broadcom, 2019).

7.2.6 Emerging Tech and Cyber

In response to the rapid growth of IoT and associated attacks, the Netherlands and Singapore, published an IOT Security Landscape study as part of the Smart Nation agenda (van Staalduinen & Joshi, 2019). As wearables increased in popularity, there were concerns about the privacy of healthcare data (Ranger, 2019). Researchers also started to consider who is liable when autonomous vehicles get hacked (Winkelman, et al., 2019).

7.2.7 Enhanced Regulatory Framework

Since 2018, Australia, the United States, and the United Kingdom have all introduced new laws and regulations to enhance the governance of cyberspace. In 2018, Australia enacted the *Security of Critical Infrastructure Act*, which establishes a register of critical infrastructure assets aimed at developing a clear picture of the ownership and control of Australia's critical infrastructure assets across high-risk sectors and supporting proactive risk-management activities (Department of Home Affairs, 2020). Furthermore, in December 2020, the United States passed the *IoT Cybersecurity Improvement Act of 2020* – in part a response to the Mirai malware, which created a botnet from IoT devices (such as security cameras, smart TVs, and other such smart-devices) to launch large-scale DDoS attacks – which outline the baseline security requirements IoT devices must have before being considered for purchase by the United States government (Murphy, 2020). Finally, data privacy has emerged as a major area of interest among regulators. In 2018, the United Kingdom enacted the *Data Protection Act, and the Data Protection, Privacy and Electronic Communications (Amendments etc.) (EU Exit) Regulations* in 2019. Both pieces of legislation work to implement the requirements from the European Union's *General Data Protection Regulation* into domestic UK law prior to Brexit (HM Government, 2019). Similarly, numerous states in the United States have enacted their own data privacy laws. These include California, New York, Maine, Oregon, Nevada, Texas, New Jersey, Washington, and Massachusetts.

7.3 MOVING INTO 2020: THE EFFECTS OF THE PANDEMIC

Within this landscape, analysts were predicting key topics for 2020 like maturing risk appetite, security operations and data security (Panetta, 2019), weaponisation of AI and Privacy class actions (Forrester, 2019b). 2019 was a year where technology marched ahead and cyber risk stayed at the top of the agenda, but nobody could actually predict what was to come with COVID-19 in 2020 and what would be left in its aftermath. It should be no surprise that a world-changing event like the latest pandemic, that disrupted the lives and works of billions of people, would offer a fertile ground for criminals to find new ways to implement their malicious schemes and attacks. Indeed, the abrupt changes that were required with remote work becoming the norm forced companies worldwide to restructure and adapt to a new environment without the necessary preparations. Companies having dozens of people working together in a single office now had to continue being productive with each employee working in physical isolation in their own home/office. Productivity became the first and foremost concern leaving everything else, including cyber security related concerns and practices, for granted. In this context, it is no surprise that the pandemic actually acted as an accelerator of threats, where cyber criminals, like the actual virus, found a very fertile environment where to multiply and spread their attacks to a multitude of new targets, unprepared to face such threats from both a technological as well as a psychological perspective.

Reports such as the McAfee Labs Threat Report (2020, p. 5) show an average of 419 new threats are launched per minute and, indeed, several areas across all industries saw an exponential growth in threats of different nature, targeting physical infrastructure and servers, corporate identity-based frauds, ransomware attacks and more, preying on the distress and vulnerability of common people to implement new scams whenever possible.

7.3.1 Home Office Is under Attack

With an unparalleled number of people forced at home, it is easy to predict how internet traffic would have spiked to unprecedented levels all over the world[1] (Yip, 2020; Koeze and Popper, 2020). This meant not only additional stress across the infrastructure but also an increasing number of devices being connected and, while companies all over the world spent a huge of resources to secure their online perimeter in their headquarters and offices, the unplanned shift to working from home did expose workers and assets to completely unexpected vulnerabilities. The 2020 Xfinity Cyber Health Report (2020, p. 10) estimated that an average household in the United States has 12 devices connected to the internet, with tech-savvy users having more than 30 devices online. With such a plethora of possible targets, it is not surprising then to realise that each home is actually subjected to more than one hundred threats each month. Most importantly, though, the report underlined how users tend to underestimate the impact and severity of threats and that any online device like laptops, PCs, smart phones, printers, networked storage devices, smart TVs etc. can be considered as a target and as a possible entry point not only to the user's own personal data but, potentially, also to sensible company data and information. In fact, while direct connections to the office network may be protected via virtual private network (VPN) and other security measures, it is likely that users may copy files on their devices and storage space. This makes phishing and ransomware threats a constant danger that can have far-reaching consequences beyond the individual user.

The mixing of working activities in a more open space like a home setting is prone to many more threats that also need to be considered. For example, while business meetings may be encrypted to keep them secure, an intruder could still listen to every detail from a hacked nearby device, like a smart phone or even the microphone built in the remote controller of a smart TV.

If all these technical threats were not enough to make the life of security professionals difficult, the most critical weakness does remain the human element and our own lack of security awareness, something that is present at all levels of the corporate ladder. This was made painfully apparent when, on November 20, 2020, a young Dutch journalist managed to hack into a confidential meeting between the EU defence ministers thanks to one of them sharing a screenshot on social media in which the pin to access the ongoing connection was visible (Deutche Welle, 2020a).

7.3.2 Attacks towards Specific Companies' Infrastructure

While targeting the new home-office environment, hackers did not forget about the proper office or industry infrastructure, though. During this time, ransomware attacks had an unprecedented rise everywhere, in some cases with lethal consequences, as shown by an attack directed to the Heinrich Heine University in Dusseldorf in Germany that locked a server in the University Hospital instead, preventing the facility to check-in a patient in critical condition (Ralston, 2020), but these were not the only focus of black hat hackers.

Sophistication of botnet-based attacks is in constant evolution and allows for a variety of applications, from DDoS attacks, to break down services and companies' operations, to stealing computing power for crypto mining purposes. These attacks are also making no differences in identifying targets and are impacting all industrial sectors.

Critical infrastructure, such as the power grid, is not only at risk of being damaged but can also be manipulated by IoT botnets, not only to cause obvious outages and blackouts but also in more subtle ways to cause electricity price fluctuations.

The manufacturing sector, including food and beverage, automobile and semiconductor industries, is constantly under pressure as hackers try to exploit every opportunity or any potential weakness to steal sensitive information: Razer, a gaming hardware manufacturer leader, suffered a data leak due to an unsecured database exposing customer data online (Lyles, 2020) and electrical car maker Tesla barely avoided potentially very serious consequences by successfully unfolding an ongoing attack to its Nevada plant, where hackers were trying to bribe an employee by offering him 1 million USD, to install their malware on some of the firm PCs (Reuters, 2020).

Finally, the retail sector is also under constant attack as any breach in a company's own users database offers valuable data and personal information that can be exploited by providing, for example, a never-ending source of possible targets for following phishing attacks.

7.3.3 Everything Is a Target: How Is This Happening?

As exemplified in the previous section, literally anything can be a target, from high profile public companies to small online businesses, from critical infrastructure to government organisations. Many different techniques can be used to exploit vulnerability and launch an attack but most of these tend to start by relying on what is considered as the weakest element of any cyber defence: the end-user, who is seen as a point of entry for deploying a specific malware. This usually happens via the so-called phishing, where an attacker tries to lure an unsuspecting user to install, or otherwise introduce in some way, a malicious programme into the target network or computers. The attack starts via some email, either generic or with content tailored to suit a specific target, to bait him or her into clicking a link from which the malware is ultimately deployed or relevant personal information, like usernames and passwords, is obtained.

In this regard, the pandemic did not offer hackers only an opportunity to expand their activities, but also an excellent topic to increase the chances of success: a pandemic and all its possible consequences and threats is a perfect fit to prey on people's fears and elicit a sense of urgency, which is commonly needed to convince someone in reading the message and clicking what would, in the end, open the door to the actual attack.

One of the most troublesome characteristics of these attacks is that they do not necessarily need special technical skills to be deployed effectively: even inexperienced attackers can be able to craft a dangerous threat. This means that even for the common user and, hence, for a work-from-home employee, it is then of paramount importance to understand how these attack work, so that we can spot hints and signs of malicious activity as early as possible.

Penetration testing tools such as the Social Engineering Toolkit[2] and ShellPhish[3] make the mechanics of a social engineering attack painfully apparent. The former allows an attacker to mimic any existing website while the latter is focused on social websites such as Instagram, Facebook, Twitter, Snapchat and GitHub. The fake websites are programmed so that, once users input their credentials, they forward these to the attacker while redirecting the users to the real website. Users will likely just think of some small connection hiccup when asked, by the real website this time, to input their credentials to login once again and comply without realising what just happened. With such set-up, the attacker needs only focus on crafting an engaging and realistic email to redirect potential victims to the laid down trap.

7.4 REASSESSING RISK

COVID-19 caused a sudden, worldwide requirement for digital transformation programmes in order to effectively facilitate remote work. Indeed, according to the Harvey Nash/KPMG CIO Survey (2020, p. 9) 'enabling the workforce' has become one of the top three issues for business during the pandemic, with new technology and tools being developed and deployed to ensure minimal productivity losses and to facilitate effective remote working regimes. Additionally, the survey found that the five most important technology investments organisations have been making during the pandemic are those to improve security and privacy, customer experience and engagement, infrastructure/cloud, automation and business intelligence.

Organisations are increasingly vulnerable as a result of technological advances and changing working practices such as remote access, big data, cloud computing, social media and mobile technology.

As cloud-based solutions have been rapidly adopted to facilitate remote work, organisations have similarly ensured the rapid adoption of additional technology to help ensure that sensitive information stored online remains secure (ibid., p. 13). This rapid digital transformation has also seen significant changes in the cyber security landscape, and the nature of cyber-attacks. Due to remote work, employees themselves are now the biggest risks to the organisation's cyber security, as is suggested by an 83%

increase in spear-phishing attacks, and 62% increase in malware during the pandemic (ibid., p. 18).

With the COVID-19 pandemic forcing the world to rapidly adopt new tools and technologies, in many cases without conducting a full security review prior to implementation; it is inevitable that many large-scale cyber-attacks will follow.

Security experts have been warning about the rapidly increasing sophistication of cyber-attacks throughout the pandemic (Microsoft, 2020, p. 18; Davis and Pipikaite, 2020), which is likely to culminate in a large-scale cyber-attack in the post-pandemic world.

Due to lockdowns and social distancing rules, we have become more digitally connected than ever before in order to continue to socialise with distant friends and family and work effectively. As such, any large-scale cyber-attack targeting software or systems widely used by public and private organisations and individuals has the potential to cause significant damage.

There is a significant responsibility on the part of boards, audit committees and executives to ensure to customers, stakeholders and regulators that appropriate cyber safeguards are in place, commensurate with the risk, nature and complexity of the business. As a result of COVID, corporates need to reassess their cyber risks.

7.4.1 Solarwinds Orion – An Image of the Future

Given the rapid digitalisation of the global workforce and growth of espionage related cyber activities, perhaps the largest and most significant cyber threat in the post-pandemic world is a single large-scale cyber-attack affecting public and private entities around the world. Indeed, such an event has been theorised (Magee, 2013; Arquilla, 2009), however in the post-COVID world such an event is increasingly likely to occur.

In late December 2020 – just as the rollout of COVID-19 vaccines were beginning in some countries – SolarWinds Orion, a system used to manage IT resources by over 300,000 clients (SolarWinds Corporation, 2020) including several US Federal Government departments, large multinational corporations, and small businesses, was hacked by a suspected state-based actor. The attack was launched in March, and involved malware being inserted into routine software updates which were then installed by up to 18,000 organisations (Satter and Bing, 2020) including several major US federal government agencies and departments as well as numerous Fortune 500 companies (FireEye, 2020a; Smith, 2020). According to former US presidential homeland security adviser, Thomas Bossert, "it is likely that the attackers will have gained 'persistent access' – the ability to manipulate, infiltrate, and control networks in a manner that is largely undetectable" (Bossert, 2020).

As we saw, there has been a significant spike in the utilisation of cyberspace to steal strategic information during the pandemic, with healthcare and research institutions working on vaccine development bearing the brunt of such attacks – the Federal Bureau of Investigation and Cybersecurity and Infrastructure Security Agency released a statement in May 2020 explicitly warning organisations working on a COVID-19 vaccine that cyber-threat actors affiliated with the Chinese government are attempting to steal their intellectual property (FBI and CISA, 2020). The rapid normalisation of remote work – Deloitte, Microsoft and the Australian Public Service, among others, are working to make remote work a more permanent fixture in the future (Mathew, 2020; Hogan, 2020; Burton, 2020; Ortega, 2020), for example – is potentially increasing the risk that such attacks will be successful, as organisations may have to compromise security in some way to ensure employees remain able to easily access sensitive information while working from home. Indeed, just over 50% of organisations are most worried about the increased risks posed by cyber-attacks purely because of the shifting working environment – behind only a likely surge in bankruptcies across industries and a global recession (World Economic Forum, 2020, p. 12).

7.4.2 Digital Transformation – New Risks

The pandemic has also seen a large surge in digital transformation across industries. Retailers, for example, have increasingly adopted technologies such as augmented reality, chatbots, various forms of cashless payments, and digital receipts in order to minimise human contact while maintaining regular business operations as much as possible (Dixon and Singh, 2020), and in some cases, enhancing the customer experience as they shop in a COVID-safe manner (KPMG Italy, 2020). Indeed, Amazon's Just Walk Out technology allows retailers to eliminate human contact and checkout queues in physical stores (Associated Press, 2021). Similar adoption of new technology to increase safety in-line with global social distancing guidelines and lockdown rules has occurred across all industries and sectors to ensure the continuation of business as much as possible.

As a result of the rapid digital transformation, numerous cyber risks that were faced by businesses prior to the pandemic have suddenly been amplified, and will pose a greater danger to organisations going forward purely due to the normalisation of remote work.

As the SolarWinds attack highlighted, vulnerabilities within third party tools can pose a significant cyber risk to organisations, and particularly to less mature small and medium enterprises. In particular, tools being utilised for communication and the use of non-approved and personal hardware and software to effectively work from home have significantly increased the cyber risks faced by all organisations, as remote working tools are often the target of cyber threat actors (International Chamber of Commerce, 2020), and as employees are increasingly likely to use personal devices for work, or corporate devices for personal use, while working from home, more activity is being conducted outside of organisations' firewalls and network perimeters. Indeed, the controls organisations have been utilising to maintain security over their digital have become increasingly ineffective. As employees moved from working on secure corporate networks to home Wi-Fi, data which was only accessible when connected to corporate networks has, too, in some cases moved to less secure digital locations to allow employees easier access to data while working at home (Galligan and Golden, 2020; Bates, 2020).

Perhaps the biggest risk that has emerged during the pandemic which will continue to be an issue is that of insider threats. As employees work from home, they are more prone to become unwitting insider threats as they are more prone to distraction and stress at home rather than in a dedicated workplace, and are much more prone to making errors (CERT Insider Threat Team, 2013, pp. 7–10), such as not connecting to a VPN, and failing to identify phishing emails. This risk, coupled with information security controls which have become weaker due to remote work, has significant potential to be exploited by cyber adversaries and threat actors, resulting in increasingly intense cyber-attacks due to the significantly increased potential attack surface resulting from the increased number of endpoints and digital technologies organisations have deployed to facilitate remote work.

While the aims and objectives pursued by adversaries operating in cyberspace has largely remained the same throughout the course of the pandemic, and will likely remain so long afterward, the rapid shift in the work environment has made it much easier for these objectives to be carried out – as Dr. Danny Steed (2020) states, "[t]he scramble to fully equip workforces with laptops, remote access and (hopefully) VPNs changed the calculus of risk for any organisation adopting these measures – they all became purely online businesses immediately". This shift in the nature of the work environment, if not handled with due care by organisations, will have significant potential to amplify the already significant risks posed by cyber threat actors. As new technologies are developed and deployed, and as new strategic information is stored in what is now a less-secure manner, the risk of a major attack resulting in IP theft and causing significant reputational damage to a targeted entity (or entities) is significantly amplified. In this sense, it is almost inevitable that an attack similar in style and scope as the SolarWinds attack will occur again in the near future as more data is hosted online to facilitate remote work in the new normal.

7.4.3 Global Politics and Cyber Security – Growing Targets for Cyber Spies

As geopolitical tensions rise, and as COVID-19 has forced more work to be done online, various countries will more than likely step up their cyber-espionage activities. Such activities can be done for numerous reasons. Perhaps one of the most significant is for intelligence gathering, as cyber allows adversaries a highly cost-effective method of stealing large amounts of strategic information while adding an extra layer of deniability. In the case of North Korea, cyber-attacks serve the additional purpose of being a way to increase the revenue of the country under significant financial sanctions.

Indeed, in 2019 it was reported that North Korea stole US$2 billion to fund its weapons programmes by conducting cyber-attacks against banks and cryptocurrency exchanges (Nichols, 2019). With the impact that the pandemic has had on the country, particularly on trade and the economy (Deutsche Welle, 2020a,b), it is highly probable that North Korean linked hacker groups will begin new campaigns targeting banks, cryptocurrency exchanges, and other financial institutions primarily based in South Korea in order to secure funding for the Kim regime during the post-pandemic recovery.

IP theft and intelligence gathering are no longer the only purposes behind state-backed offensive cyber operations, however. Increasingly, we have seen cyber-attacks attributed to state-actors which have been designed to act as a method of political coercion. In this sense, cyber security is no longer a purely technical issue, but an increasingly geopolitical one. Utilising cyberspace allows states and non-state actors to aggressively pursue their strategic objectives, exert influence over other countries, and force opposing state and non-state actors to consider the costs of continued opposition (Flemming and Rowe, 2015, pp. 97–99; Hodgson, 2018, pp. 74–76). At particular risk are organisations, such as SolarWinds, which develop tools and technology that are widely used across various sectors.

7.4.4 No Longer Limited to Cyberspace – The Crippling Effects of Cyber-Attacks

As the world has become increasingly dependent on interconnected technology solutions, with this dependency seeing a sharp spike during the course of the pandemic, and with the developments in offensive cyber capabilities from various threat actors in recent years, there is a significant risk for cyber-attacks to cause significant physical and economic damage. Indeed, as the CEO of IT security firm Check Point stated, the global pandemic has "...pushed forward five, maybe even 10 years of technological evolution," making it easier for crown jewels to be accessed (Reich, 2020).

With the amount of technological connections that exist today, it is probable that malware similar to Mirai, which built a botnet using IoT devices to launch what were the biggest DDoS attacks ever recorded at the time, will be used to launch attacks similar to what was experienced in Estonia in 2007, when the country was 'besieged' by 128 DDoS attacks over two weeks, shutting down all banks, ATMs, telecommunications, and media outlets (Scharre, 2018, p. 239). Attacks causing similarly physical damage have occurred recently; however, none have been on the same level as that which was seen in Estonia.

Attacks targeting power stations, for example, potentially cutting off the electricity supply to large portions of cities, have occurred – the first-known successful example of this was following the Ukrainian crisis of 2013–2014 and subsequent revolution, where Russian-backed attackers were able to shut off the power in parts of Ukraine's Ivano-Frankivsk region, affecting roughly 225,000 residents for up to 6 hours in the middle of winter (CISA, 2018; Zetter, 2016).

Furthermore, as has been mentioned previously in this chapter, the University Hospital Düsseldorf suffered a major ransomware attack, forcing the hospital to operate at half capacity and forcing ambulances to divert – playing role in the death of a patient who had their treatment delayed (Eddy & Perlroth, 2020).

The commonality in these attacks is the human element. Spear-phishing emails were used to insert malware into the Ukrainian power grid (Zetter, 2016), and a vulnerability in the Citrix VPN used by the University Hospital Düsseldorf which Germany's Cyber Security Council warned of in January was left unpatched (Tidy, 2020). With attacks on critical infrastructure – such as hospitals and power grids – likely to see a significant increase in intensity in the near future, it is paramount that organisations recognise and adequately address the human element of cyber-attacks. With remote work showing no signs of going away, this will be the key factor that attackers will focus on exploiting in the post-pandemic world.

7.5 CONCLUSION – WHAT CAN BE DONE?

A lot happened during and after the COVID-19 pandemic and an already complex and challenging cyber-threat scenario quickly evolved in different ways, including some unexpected directions where anything and anybody, from a single individual in a small business to multinational corporations and even entire nations can be targeted and compromised. In order to mitigate the threat posed by large-scale cyber-attacks and cyber-espionage activities, organisations should adopt a cyber-risk framework.

But how to actually manage cyber-risk? For this, we should consider six key pillars: leadership and governance, human factors, information risk management, technology and operations, business continuity management, and legal and compliance. An overarching cyber strategy should align the cyber programme with the business needs and links the six pillars together.

Leadership and governance identifies the board's and leadership's understanding of Cyber and its Cyber-risk appetite, and demonstrating due diligence, ownership and effective management of risk. Key challenges include under-staffed security organisations, taking an IT-centric approach and a lack of alignment with the business needs. Corporates should assign a Chief Information Security Officer (CISO), define their risk appetite, implement security policies and report cyber-risk regularly to the board.

Human factors ensure the organisation's personnel are properly trained and understand the nature of cyber security and their role in protecting the organisation's assets. Key challenges include maintaining engaging awareness sessions and developing training for specific groups. Corporates should conduct role-based security awareness programmes for new and existing staff and assess its effectiveness through phishing and social engineering exercises. Background screening should also be implemented that is appropriate to job function and/or authority level.

Information risk management focuses on an effective risk management framework to manage information risk throughout the organisation and with its delivery and supply partners. Key challenges include not having a list of critical assets across the business, low-quality cyber-risk assessments and lack of visibility for stakeholders. Corporates should implement a data classification policy, conduct a cyber-risk assessment, maintain a cyber-risk register and report regularly to the board.

Technology and operations focuses on technical and operational control measures implemented to address identified risks and help minimise the impact of compromise. Key challenges include poorly defined processes, misconfigured tools and security not being included in new products and services. Corporates should implement baseline IT security standards and implement key capabilities including IT asset management, user and privileged access management, vulnerability and patch management, anti-virus management, secure system development lifecycle, security logging and monitoring, incident management, backups and physical security.

Business continuity management ensures that the role of security and its impact on business continuity is fully understood and integrated into the overall business continuity, crisis communications and mass notification processes. Key challenges include cyber scenarios not being identified as part of the business continuity plan and cyber incident response plans not being detailed enough. Corporates should define and test cybersecurity scenarios as part of their business continuity plans and exercises.

Legal and compliance ensures the organisation is aware of its legal and compliance obligations relating to cyber security and complies with them in an effective manner. Key challenges include identifying cyber-related regulations in multiple countries and maintaining ongoing compliance. Corporates should identify all relevant legal and regulatory obligations on cyber, resiliency, third party and privacy and ensure implementation of key controls.

If corporates adopt a cyber-risk framework they will be better placed to prevent, detect, respond and recover from cyber-attacks.

To address some of the COVID-related cyber risks, corporates will need to focus on enhancing remote access security, as well as put an increased focus on mitigating social engineering attempts.

In order to ensure cyber security effectively scales with increased remote working, there are multiple actions for organisations to take. Perhaps one of the biggest issues with remote working is maintaining access control and proper authentication. As such, organisations should ensure that they have implemented multi-factor authentication when remote access is required to organisational systems. Additionally, organisations should maintain conditional access, or implement a cloud access security broker (CASB) that is hosted on the cloud to ensures IT security policies are enforced no matter what device is attempting to access cloud-hosted data (Ferbrache, 2020). Access to the organisation's network should also be strictly controlled using a VPN, accessed by using different login credentials to those used to login to other organisational portals and systems. While the risks of the utilisation of personal devices can be mitigated with a CASB, it does not eliminate the risks of shadow IT and the use of personal devices for work. As the lines between the 'home' and 'office' get blurred with work from home, the use of non-approved software and hardware will significantly increase. As such, organisations need to work to help ensure that approved tools and solutions are convenient and easy to use, while also being open to feedback regarding the adoption of new solutions and tools in order to minimise the shadow IT risks (ibid.).

Finally, and perhaps most importantly, organisations must rework their cyber training and awareness programmes to provide additional guidance on the cyber risks specific to remote working and increase emphasis on aspects concerning the human element of cyber-attacks.

Prior to the COVID-19 pandemic, the human element was always been the weak link in in cyber security. It is humans, after all, that can be duped into downloading malware or entering their login credentials into seemingly legitimate websites. With the rise of remote work and rapid acceleration of various digital transformation programmes to meet the new requirements organisations were forced into by global lockdowns and social distancing rules, the human aspect of cyber security is even more important.

Unfortunately, we can easily predict that more challenges are ahead: we were given a taste of what is inevitably on the horizon for cyber security with the cyber-attack on SolarWinds Orion, which had a global impact. Such large-scale attacks will likely become commonplace as the surface area of attack for many organisations will have significantly expanded as a result of the aforementioned remote work and digital transformation, which, alongside its many benefits, brings in also additional, and possibly lethal, dangers.

NOTES

1. See https://www2.telegeography.com/network-impact for the latest collated reports and news.
2. See https://www.trustedsec.com/tools/the-social-engineer-toolkit-set/.
3. See https://github.com/suljot/shellphish.

REFERENCES

Accenture Security. (2019). *Cyber Threat Landscape Report*. Retrieved from https://www.accenture.com/_acnmedia/PDF-107/Accenture-security-cyber.pdf

Arquilla, J. (2009, July 26). Click, click... counting down to Cyber 9/11. *San Francisco Chronicle*, E2.

Associated Press. (2021, January 12). *Hudson Advances Digital Growth Strategy With Plans to Open Stores Using Amazon's Just Walk Out Technology*. Retrieved from Associated Press: https://apnews.com/press-release/business-wire/technology-lifestyle-travel-business-new-york-ca9c00ea22da4e288932341e046219b1

Bates, S. (2020, March). *Managing the information security impact of COVID-19 - Risk and security threats related to remote working*. Retrieved from KPMG International: https://home.kpmg/xx/en/home/insights/2020/04/managing-the-information-security-impact-of-covid-19.html

Bossert, T. P. (2020, December 16). *I was the Homeland Security Adviser to Trump. We're Being Hacked*. Retrieved December 23, 2020, from The New York Times: https://www.nytimes.com/2020/12/16/opinion/fireeye-solarwinds-russia-hack.html

Broadcom. (2019, November 04). *Broadcom Completes Acquisition of Symantec Enterprise Security Business*. Retrieved from Broadcom: https://investors.broadcom.com/node/52706/pdf

Burton, T. (2020, May 13). *Public servants may stay home permanently*. Retrieved December 24, 2020, from Australian Financial Review: https://www.afr.com/politics/federal/public-servants-may-stay-home-permanently-20200513-p54sjg

Center for Strategic and International Studies. (2021). *Significant Cyber Incidents Since 2006*. Washington D.C.: Center for Strategic and International Studies. Retrieved from https://csis-website-prod.s3.amazonaws.com/s3fs-public/201218_Significant_Cyber_Events.pdf

CERT Insider Threat Team. (2013). Unintentional Insider Threats. Pittsburgh: Carnegie Mellon University, Software Engineering Institute.

CISA. (2018, August 23). *ICS Alert (IR-ALERT-H-16-056-01) – Cyber-Attack Against Ukrainian Critical Infrastructure*. Retrieved December 27, 2020, from Cybersecurity and Infrastructure Security Agency: https://us-cert.cisa.gov/ics/alerts/IR-ALERT-H-16-056-01

Cisco. (2019). *Accepting the Unknowns – Chief Information Security Officer Benchmark Study*. San Jose: Cisco Systems Inc.

Clement, J. (2020, May 27). *IC3: total damage caused by reported cyber crime 2001–2019*. Retrieved from *Statista*: https://www.statista.com/statistics/267132/total-damage-caused-by-by-cyber-crime-in-the-us/

Continuity Central. (2019a, January 09). *Three trends that will impact disaster recovery in 2019*. Retrieved from Continuity Central: https://continuitycentral.com/index.php/news/technology/3610-three-trends-that-will-impact-disaster-recovery-in-2019

Continuity Central. (2019b, January 18). *Business continuity trends and challenges 2019: survey results*. Retrieved from Continuity Central: https://www.

continuitycentral.com/index.php/news/business-continuity-news/3643-business-continuity-trends-and-challenges-2019-survey-results

Council on Foreign Relations. (2021). *Cyber Operations Tracker*. Retrieved from Council on Foreign Relations: https://www.cfr.org/cyber-operations/

Davis, N., & Pipikaite, A. (2020, June 01). *What the COVID-19 pandemic teaches us about cybersecurity – and how to prepare for the inevitable global cyberattack*. Retrieved December 25, 2020, from World Economic Forum: https://www.weforum.org/agenda/2020/06/COVID-19-pandemic-teaches-us-about-cybersecurity-cyberattack-cyber-pandemic-risk-virus/

Department of Home Affairs. (2020, December 16). *Security coordination - Security of Critical Infrastructure Act 2018*. Retrieved from Department of Home Affairs: https://www.homeaffairs.gov.au/about-us/our-portfolios/national-security/security-coordination/security-of-critical-infrastructure-act-2018

Department of Justice. (2019, December 5). *Russian National Charged with Decade-Long Series of Hacking and Bank Fraud Offenses Resulting in Tens of Millions in Losses and Second Russian National Charged with Involvement in Deployment of "Bugat" Malware*. Retrieved from Department of Justice: https://www.justice.gov/opa/pr/russian-national-charged-decade-long-series-hacking-and-bank-fraud-offenses-resulting-tens

Deutche Welle. (2020a, November 20). Dutch reporter hacks EU defense ministers' meeting. Retrieved from Deutche Welle: https://www.dw.com/en/dutch-reporter-hacks-eu-defense-ministers-meeting/a-55682752

Deutsche Welle. (2020b, June 09). *'Some starving' in North Korea as COVID constrains China trade, say UN experts*. Retrieved December 27, 2020, from Deutsche Welle: https://p.dw.com/p/3dXyI

Dignan, L. (2019a, August 15). *Top cloud providers 2019: AWS, Microsoft Azure, Google Cloud; IBM makes hybrid move; Salesforce dominates SaaS*. Retrieved from ZDNet: https://www.zdnet.com/article/top-cloud-providers-2019-aws-microsoft-azure-google-cloud-ibm-makes-hybrid-move-salesforce-dominates-saas/

Dignan, L. (2019b, March 20). *Global security spending to top $103 billion in 2019, says IDC*. Retrieved from ZDNet: https://www.zdnet.com/article/global-security-spending-to-top-103-billion-in-2019-says-idc/

Dixon, W., & Singh, M. (2020, July 27). *COVID-19 has disrupted cybersecurity, too – here's how businesses can decrease their risk*. Retrieved from World Economic Forum: https://www.weforum.org/agenda/2020/07/covid-19-cybersecurity-disruption-cyber-risk-cyberattack-business-digital-transformation/

Eddy, M., & Perlroth, N. (2020, September 18). *Cyber Attack Suspected in German Woman's Death*. Retrieved December 27, 2020, from The New York Times: https://www.nytimes.com/2020/09/18/world/europe/cyber-attack-germany-ransomeware-death.html

Evans, D. (2011). *The Internet of Things: How the Next Evolution of the Internet is Changing Everything*. San Jose: Cisco Internet Business Solutions Group. Retrieved from https://www.cisco.com/c/dam/en_us/about/ac79/docs/innov/IoT_IBSG_0411FINAL.pdf

FBI & CISA. (2020, May 13). *People's Republic of China (PRC) Targeting of COVID-19 Research Organizations*. Retrieved December 26, 2020, from Cybersecurity and Infrastructure Security Agency: https://www.cisa.gov/sites/default/files/publications/Joint_FBI-CISA_PSA_PRC_Targeting_of_COVID-19_Research_Organizations_S508C.pdf.pdf

Ferbrache, D. (2020, April 17). *Scaling security for remote working*. Retrieved December 26, 2020, from KPMG International: https://home.kpmg/xx/en/home/insights/2020/04/scaling-security-for-remote-working.html

FireEye. (2020a, December 13). *Threat Research – Highly Evasive Attacker Leverages SolarWinds Supply Chain to Compromise Multiple Global Victims With SUNBURST Backdoor*. Retrieved December 23, 2020, from FireEye Inc.: https://www.fireeye.com/blog/threat-research/2020/12/evasive-attacker-leverages-solarwinds-supply-chain-compromises-with-sunburst-backdoor.html

FireEye. (2020b). *Advanced Persistent Threat Groups – Who's who of cyber threat actors*. Retrieved December 27, 2020, from FireEye: https://www.fireeye.com/current-threats/apt-groups.html

Fitch Ratings. (2020, April 29). *US Cyber Insurance Market Will Be Tested by Coronavirus Fallout*. Retrieved from Fitch Ratings: https://www.fitchratings.com/research/insurance/us-cyber-insurance-market-will-be-tested-by-coronavirus-fallout-29-04-2020

Flemming, D. R., & Rowe, N. C. (2015). Cyber Coercion: Cyber Operations Short of Cyberwar. *Proceedings of the 10th International Conference on Cyberwarfare and Security ICCWS-2015, Skukuza, South Africa, March*, 95–101.

Forrester. (2019a, December 18). *Decade Retrospective: Cybersecurity From 2010 To 2019*. Retrieved from Forbes: https://www.forbes.com/sites/forrester/2019/12/18/decade-retrospective-cybersecurity-from-2010-to-2019/?sh=498931b44d51

Forrester. (2019b). *Predictions 2020 - On The Precipice Of Far-Reaching Change*. Cambridge, MA: Forrester Research.

Galligan, M., & Golden, D. (2020, November 23). *Cyber: New Challenges in a COVID-19–Disrupted World*. Retrieved from Harvard Law School Forum on Corporate Governance: https://corpgov.law.harvard.edu/2020/11/23/cyber-new-challenges-in-a-covid-19-disrupted-world/

Georgia Technology Authority. (2019). *Georgia Cyber Center*. Retrieved from Georgia Technology Authority: https://gta.georgia.gov/cybersecurity-1/georgia-cyber-center

Harvey Nash & KPMG. (2020). *Harvey Nash/KPMG CIO Survey 2020*. London: Harvey Nash.

HM Government. (2019, March 07). *Explanatory Memorandum to The Data Protection, Privacy and Electronic Communications (Amendments etc) (EU Exit) (No.2) Regulations 2019*. Retrieved from GOV.UK: https://assets.publishing.service.gov.uk/media/5c6bde6ded915d4a343cb9df/EM_to_The_Data_Protection__Privacy_and_Electronic_Communications.pdf

Hodgson, Q. E. (2018). Understanding and Countering Cyber Coercion. *10th International Conference on Cyber Conflict (CyCon).*, 73–88.

Hogan, K. (2020, October 9). *Embracing a flexible workplace.* Retrieved December 24, 2020, from Microsoft: https://blogs.microsoft.com/blog/2020/10/09/ embracing-a-flexible-workplace/

International Chamber of Commerce. (2020). *COVID-19 Cyber Security Threats to MSMEs.* Retrieved from International Chamber of Commerce: https:// iccwbo.org/content/uploads/sites/3/2020/05/2020-icc-sos-cybersecurity.pdf

(ISC)². (2019a, November 06). *(ISC)² Finds the Cybersecurity Workforce Needs to Grow 145% to Close Skills Gap and Better Defend Organizations Worldwide.* Retrieved from (ISC)²: https://www.isc2.org/News-and-Events/Press-Room/ Posts/2019/11/06/ISC2-Finds-the-Cybersecurity-Workforce-Needs-to-Grow--145

(ISC)². (2019b). *(ISC)² Cybersecurity Workforce Study: Women in Cybersecurity.* Clearwater: (ISC)². Retrieved from https://www.isc2.org/research/-/media/ 67B23A98D3A54E878FF927748D8F3EF1.ashx

Koeze, E., & Popper, N. (2020, April 07). *The Virus Changed the Way We Internet.* Retrieved from *The New York Times:* https://www.nytimes.com/ interactive/2020/04/07/technology/coronavirus-internet-use.html

KPMG International. (2019). *Agile or Irrelevant: Redefining resilience - 2019 Global CEO Outlook.* Amstelveen: KPMG International. Retrieved from KPMG International: https://assets.kpmg/content/dam/kpmg/xx/pdf/2019/05/kpmg-global-ceo-outlook-2019.pdf

KPMG Italy. (2020, April). *COVID-19 and social distancing impact on Retail Customer Experience: KPMG vision & approach for Large Retail Chains.* Retrieved from KPMG International: https://assets.kpmg/content/dam/kpmg/ it/pdf/2020/04/KPMG_COVID-19_New-Retail-Customer-Experience.pdf

KPMG US. (2019, March). *Top cyber security considerations in 2019.* Retrieved from KPMG US: https://boardleadership.kpmg.us/relevant-topics/articles/ 2019/cyber-security-considerations-for-2019.html

Lyles, T. (2020, September 14). *Razer accidentally leaked the personal information for over 100,000 gamers, report says.* Retrieved from The Verge: https:// www.theverge.com/2020/9/14/21436160/razer-data-leak-elasticsearch-sever-misconfiguration

Magee, C. S. (2013). Awaiting Cyber 9/11. *Joint Force Quarterly, 3rd Quarter,*(70), 76–82.

Mathew, R. (2020, October 19). *Corporate giant Deloitte is permanently closing 4 offices, putting 500 staff on work-from-home contracts.* Retrieved December 24, 2020, from Business Insider: https://www.businessinsider. com/deloitte-offices-covid-work-from-home-remote-working-2020-10

McAfee Labs. (2020). *McAfee Labs Threats Report.* San Jose: McAfee.

Microsoft. (2020). *Microsoft Digital Defense Report.* Redmond: Microsoft. Retrieved from https://www.microsoft.com/en-us/security/business/security-intelligence-report

Morgan, S. (2019a). *2019 Official Annual Cybercrime Report.* Toronto: Herjavec Group. Retrieved from https://www.herjavecgroup.com/wp-content/uploads/2018/12/CV-HG-2019-Official-Annual-Cybercrime-Report.pdf

Morgan, S. (2019b, February 06). *2019/2020 Cybersecurity Almanac: 100 Facts, Figures, Predictions And Statistics.* Retrieved from Cybersecurity Ventures: https://cybersecurityventures.com/cybersecurity-almanac-2019/

Murphy, G. (2020, November 11). *The IoT Cybersecurity Improvement Act: Combining Tech With Policy To Address Threats.* Retrieved from Forbes: https://www.forbes.com/sites/forbestechcouncil/2020/11/11/the-iot-cybersecurity-improvement-act-combining-tech-with-policy-to-address-threats/?sh=3788039d6129

Ng, A. (2019, December 05). *Ransomware froze more cities in 2019. Next year is a toss-up.* Retrieved from CNet: https://www.cnet.com/news/ransomware-devastated-cities-in-2019-officials-hope-to-stop-a-repeat-in-2020/

Nichols, M. (2019, August 06). *North Korea took $2 billion in cyberattacks to fund weapons program: U.N. report.* Retrieved December 27, 2020, from Reuters: https://www.reuters.com/article/us-northkorea-cyber-un-idUSKCN1UV1ZX

Office of Management and Budget. (2020). *A Budget for America's Future – Analytical Perspectives.* Washington D.C.: United States Government Publishing Office. Retrieved from govinfo: https://www.govinfo.gov/content/pkg/BUDGET-2021-PER/pdf/BUDGET-2021-PER.pdf

Ortega, A. (2020, June 26). *Silicon Valley, in remote mode.* Retrieved December 26, 2020, from Elcano Royal Institute for International and Strategic Studies: https://blog.realinstitutoelcano.org/en/silicon-valley-in-remote-mode/

Osborne, C. (2019, December 12). *These are the worst hacks, cyberattacks, and data breaches of 2019.* Retrieved from ZDNet: https://www.zdnet.com/article/these-are-the-worst-hacks-cyberattacks-and-data-breaches-of-2019/

Panetta, K. (2019, June 19). *Gartner Top 7 Security and Risk Trends for 2019.* Retrieved from Gartner: https://www.gartner.com/smarterwithgartner/gartner-top-7-security-and-risk-trends-for-2019/

Ralston, W. (2020, November 11). *The untold story of a cyberattack, a hospital and a dying woman.* Retrieved December 28, 2020, from Wired: https://www.wired.co.uk/article/ransomware-hospital-death-germany

Ranger, S. (2019, January 14). *Why your smartwatch and wearable devices are the next big privacy nightmare.* Retrieved from ZDNet: https://www.zdnet.com/article/smartwatch-data-collection-rush-raises-privacy-backlash-fears/

Reich, A. (2020, June 01). *Check Point CEO: We need to prepare for the coming 'cyber pandemic'.* Retrieved from Jerusalem Post: https://www.jpost.com/jpost-tech/check-point-ceo-we-need-to-prepare-for-the-coming-cyber-pandemic-629933

Reuters. (2020, August 28). *Musk confirms Tesla Nevada factory was target of 'serious' cyberattack.* Retrieved from Reuters: https://www.reuters.com/article/us-tesla-cyber/musk-confirms-tesla-nevada-factory-was-target-of-serious-cyberattack-idUSKBN25O07K

SANS Security Awareness. (2019). *2019 Security Awareness Report: The Rising Era of Awareness Training.* Rockville: SANS Institute.

Satter, R., & Bing, C. (2020, December 16). *Hackers at center of sprawling spy campaign turned SolarWinds' dominance against it.* Retrieved December 23, 2020, from Reuters: https://www.reuters.com/article/global-cyber-solarwinds/hackers-at-center-of-sprawling-spy-campaign-turned-solarwinds-dominance-against-it-idUKL1N2IV1UQ

Scharre, P. (2018). *Army of None – Autonomous Weapons and the Future of War.* New York: W. W. Norton & Company.

Shani, T. (2019, April 30). *This DDoS Attack Unleashed the Most Packets Per Second Ever. Here's Why That's Important.* Retrieved from Imperva: https://www.imperva.com/blog/this-ddos-attack-unleashed-the-most-packets-per-second-ever-heres-why-thats-important/

Small, H., & De Fonseka, J. E. (2019, June 04). *GDPR: One Year On.* Retrieved from Baker McKenzie: https://www.bakermckenzie.com/en/insight/publications/2019/06/gdpr-one-year-on

Smith, B. (2020, December 17). *A moment of reckoning: the need for a strong and global cybersecurity response.* Retrieved December 23, 2020, from Microsoft – Microsoft On the Issues: https://blogs.microsoft.com/on-the-issues/2020/12/17/cyberattacks-cybersecurity-solarwinds-fireeye/

SolarWinds Corporation. (2020, December 14). *SolarWinds Corporation Form 8-K - Current Report.* Retrieved from United States Securities and Exchange Commission: https://www.sec.gov/ix?doc=/Archives/edgar/data/1739942/000162828020017451/swi-20201214.htm

Steed, D. (2020, July 14). *COVID-19: reaffirming cyber as a 21st century geopolitical battleground.* Retrieved December 25, 2020, from Elcano Royal Institute for International and Strategic Studies: http://www.realinstitutoelcano.org/wps/portal/rielcano_en/contenido?WCM_GLOBAL_CONTEXT=/elcano/elcano_in/zonas_in/ari94-2020-steed-COVID-19-reaffirming-cyber-as-21st-century-geopolitical-battleground

Tabrizi, B., Lam, E., Girard, K., & Irvin, V. (2019, March 13). *Digital Transformation Is Not About Technology.* Retrieved from *Harvard Business Review*: https://hbr.org/2019/03/digital-transformation-is-not-about-technology

Tidy, J. (2020, September 18). *Police launch homicide inquiry after German hospital hack.* Retrieved December 28, 2020, from BBC News: https://www.bbc.com/news/technology-54204356

United States of America v V. B. Netyksho et. al., 1:18-cr-00215-ABJ (United States District Court for the District of Columbia July 13, 2018). Retrieved from https://www.justice.gov/file/1080281/download

van Staalduinen, M., & Joshi, Y. (2019). *Internet of Things Security Landscape Study.* Singapore: Singapore Cyber Security Agency & Netherlands National Cyber Security Centre.

Verizon. (2019). 2019 Data Breach Investigations Report. New York: Verizon Communications Inc.

von Gravrock, E. (2019, March 04). *Here are the biggest cybercrime trends of 2019*. Retrieved from World Economic Forum: https://www.weforum.org/agenda/2019/03/here-are-the-biggest-cybercrime-trends-of-2019/

Winkelman, Z., Buenaventura, M., Anderson, J. M., Beyene, N. M., Katkar, P., & Baumann, G. C. (2019). *When Autonomous Vehicles Are Hacked, Who Is Liable?* Santa Monica: RAND Corporation. Retrieved from https://www.rand.org/pubs/research_reports/RR2654.html

World Economic Forum. (2019). *Regional Risks for Doing Business*. Geneva: World Economic Forum. Retrieved from http://www3.weforum.org/docs/WEF_Regional_Risks_Doing_Business_report_2019.pdf

World Economic Forum. (2020). *COVID-19 Risks Outlook: A Preliminary Mapping and Its Implications*. Geneva: World Economic Forum. Retrieved from http://www3.weforum.org/docs/WEF_COVID_19_Risks_Outlook_Special_Edition_Pages.pdf

Xfinity. (2020). *2020 Xfinity Cyber Health Report*. Philidelphia: Xfinity. Retrieved from https://update.comcast.com/wp-content/uploads/sites/33/dlm_uploads/2020/11/Cyber-Security-Booklet_Digital_Final3.pdf

Yip, W. Y. (2020, April 03). *Coronavirus: Internet data traffic spikes in S'pore as more work from home*. Retrieved from The Straits Times: https://www.straitstimes.com/tech/internet-data-traffic-spikes-in-spore-as-more-work-from-home

Zetter, K. (2016, March 03). *Inside the Cunning, Unprecedented Hack of Ukraine's Power Grid*. Retrieved December 27, 2020, from Wired: https://www.wired.com/2016/03/inside-cunning-unprecedented-hack-ukraines-power-grid/

CHAPTER 8

Reducing Cyber Risk in Remote Working

Vihangi Vagal[1] and Roberto Dillon[2]

[1]*Deloitte, London, United Kingdom*
[2]*School of Science and Technology, James Cook University Singapore, Singapore*

CONTENTS

8.1 INTRODUCTION

COVID-19 pandemic has had a widespread and devastating economic impact globally. Companies have struggled to establish and adopt new strategies to survive during the epidemic, resulting in massive financial losses. The lack of finance and resources to develop a business continuity plan and a disaster recovery plan has made it challenging for businesses to sustain and retain their operations. Remote working has positively impacted and altered the outlook of companies on working sustainably

DOI: 10.1201/9781003148715-8

over the long term. Nonetheless, remote working is dependent on factors such as secure work equipment and secure home internet infrastructure.

Whilst working from the office, companies are used to monitor, evaluate and form strategies to reduce the cyber risk by placing firewalls, intrusion prevention and detection systems as well as other methods. In many cases, although far from perfect, the continuous improvement that companies were painstakingly doing, step after step, to secure their digital assets was starting to bear fruit. And then, suddenly, the need to shift to remote working has completely re-booted the digital landscape, bringing in new challenges as well as old foes that were slowly becoming less harmful. Remote working is highly dependent on the use of reliable and secure digital technology to conduct business effectively and hence increases the possibility for massive methodical cyber loss.

The cyber threats faced by companies involve spear phishing, ransomware attacks, exposure of sensitive information via social engineering, denial of service attacks, exploiting bugs in the virtual private network (VPN) and other such threat actors. Companies may lack the support of cyber experts to overcome these threats on such a new, much larger scale. Organisations need to devise new approaches to reduce cyber risk while working from home.

Cyber security professionals are reliant on their knowledge and skills to support the organisation in protecting the confidentiality, integrity and availability of its resources. This chapter highlights the factors impacting the cyber risk associated with remote working and various strategies to reduce it.

8.2 CYBER RISK

Cyber risk has been an ongoing and growing concern globally, even before the pandemic occurred. While nations strengthened their physical security, attackers discovered innovative methods to evade defences by exploiting vulnerabilities in a computer system via technical means or social engineering, marking the beginning of cyber warfare.

In the early days of cyber warfare, attackers were relatively less skilled. Less effort was required to craft attacks that were able to compromise systems leading to loss of confidentiality, integrity and availability of sensitive information. The companies responded to these attacks by providing patches and fixes for the compromised systems. Gradually, each attack got more sophisticated and so did the defense.

The attackers began by using ingenious techniques such as social engineering to gain sensitive information. For example by pretending to be authorised personnel working for the company. In 1980, an American hacker, Kevin Mitnick compromised 40 companies, including Motorola, Nokia, Pacific Bell and other multinational corporations by mastering the art of social engineering. He carefully researched the company he was about to compromise, pretended to be authorised personnel by using the legitimate name of an employee and thus managed to acquire a series of passwords. He then compromised and gained privileged access to the system using these passwords [1].

Today, cyber-attacks impact several industries ranging from nuclear power plants, energy, healthcare, retail and wholesale, manufacturing, infrastructure, financial institutions, automation and many more. The impact not only leads to loss of reputation but also significant economic damage.

In 2015, a telecommunication company in the United Kingdom, TalkTalk was a victim of a significant data breach, where the attacker stole the personal data of about 157,000 customers. Customers of TalkTalk were contacted by unknown users asking them for their bank account details and told victims that the company had accidentally topped their account with more money. When the customer's tried to confirm the identity of the attacker, he (she) provided the customer with legitimate details such as their home address, PIN and other confidential information. The victim assumed he(she) was being a loyal customer by transferring back the money but instead anonymous transactions were made from their accounts [2, 3].

While cyber-attacks caused financial loss, they also induced physical damage to property and risk to human lives. In 2003, 'The Slammer' worm compromised the safety monitoring system and process control network at the Davis-Besse nuclear power plant in Ohio. In 2008, a 14-year-old boy in Lodz accidentally caused 4 trams to derail and 12 people were injured, when he hijacked the railways signalling system.

In 2010, the infamous worm Stuxnet, build around several zero-day exploits, infected computer systems globally and was considered as the most sophisticated malware ever developed till then. While no ownership of the attack was ever claimed, the malicious worm is generally regarded as a covert operation lead by intelligence agencies in the United States and Israel to disrupt the uranium enrichment facility, Natanz, in Iran. In fact, the malicious worm only executed itself on systems which contained specific Siemens software used to control the equipment in Natanz and spread to other computers on the local intranet looking for suitable targets to damage. Stuxnet remains an extremely dangerous example, because other attackers could replicate and modify it to infect other systems containing other software. Stuxnet marked a milestone in cyber-attacks because it showed the world how to possibly infect and damage carefully targeted cyber-physical systems [4].

In 2017, the WannaCry ransomware worm was released soon after the Shadow Brokers[1] published the exploit Eternal Blue. Eternal Blue compromised Samba shares (file sharing system) on Windows machines. The WannaCry ransomware worm was combined with Eternal Blue to compromise unpatched Microsoft Windows XP operating systems and encrypted critical information on computer systems of several companies. The victims had to pay a ransom in bitcoins to gain a key and decrypt their data. Unfortunately, the malware did not contain a method to generate a key to decrypt the information, so the victims paid the ransom but never got the decryption key. Ransomware attacks have surged in the past few years and have significantly increased during the pandemic [5].

With the increasing risks of cyber-attacks, cyber security professionals get fixated on the dangers of prevailing technology. In ref. [6], the researcher stated "We probe systems for weaknesses, scour code for flaws and knock on firewalls until they crumble. And yet, the weakest part of any system continues to be the people who use it". To overcome cyber-attacks, the companies need to secure their supply chain and most importantly spread awareness among employees. With ransomware being one of the most critical threats today affecting both companies and individuals, we should discuss its evolution and impact further.

8.3 "YOUR FILES ARE ENCRYPTED": A SHORT HISTORY OF RANSOMWARE

Ransomware is a specific type of malware that encrypts sensitive data on the victim's computer system and then offers to provide the victim with a decryption key once the ransom is paid. The sensitive data may include personal information such as photos, passwords, documents, backups and more. This definition is often part of an explanatory message that appears on the infected computer screen to inform the user of what just happened[2]. Although a victim may pay the ransom, it is not guaranteed that the user would actually receive the decryption key or tool to recover their encrypted data.

In 1989, the release of 'PC Cyborg' marked the beginning of ransomware attacks. The PC Cyborg, also commonly referred to as 'AIDS Trojan', was created by biologist Joseph Popp who handed out approximately 20,000 infected floppy disks to attendees at the World Health Organization's AIDS Conference. The trojan was dormant on the systems and remained undetected for a while but, when it activated itself, it encrypted files on the C drive using symmetric cryptography followed by a message claiming that the victims had breached a licensing agreement requiring them to pay $189 for license renewals and decryption of the disk via a cheque posted to Panama [7, 8].

Although in the last two decades of the past century hackers were, for the most part, enthusiastic hobbyists attempting to exploit systems to prove their technical skills, in the early 2000s hackers began harvesting financial gains from these cyber-attacks. As ransomware attacks evolved alongside an exponential internet penetration, the most prominent manner of distribution quickly became web-based, including infected websites and spam email attachments sent through phishing campaigns.

Notable examples of ransomware that spread in the past 20 years include 'PGPCoder' (also referred to as GPCode) that was first spotted in Russia in 2004 and spread through malicious websites, pioneering a technique referred to as 'drive-by download'. PGPCoder was a file-encrypting ransomware, or crypto-ransomware, that was active between 2005 and 2008 [9]. In 2012, Reveton worm, a screen locker ransomware, unleashed in Europe and locked the victim's computer screen. The 'Blackhole' exploit kit was used to install 'Citadel' malware on the targeted system, which then stole user passwords and credentials. The Citadel malware downloaded the ransomware, which displayed a socially engineered message indicating it was from the official federal agency, asking for a ransom amount to be transferred [10]. In 2014, Virlock, a screen locker ransomware, infected files on the system spreading through network and cloud storages and like Reveton, it also displayed a socially engineering message making victims believe it was from the official federal agency to urge affected users to pay [11].

Ransomware attacks quickly evolved from simple screen locker to more and more complex crypto-ransomware attacks. 'Cryptolocker', detected in 2013, was the trendsetter for other crypto-ransomware attacks such as Cryptowall, Cerber and Locky. It was a Trojan targeting windows users and spread throughout Australia and Europe through spam email attachments. The email attachment comprised of the hidden executable which created a window and activated a downloader, infecting the computer. It encrypted all the files on the system but couldn't self-replicate. Instead, the attackers used the Gameover ZeuS botnet (a network of infected computers remotely controlled by the botnets operator) to spread the ransomware. The attackers made about $3 million from the ransom paid by the victims [12, 13].

In late 2013, a perfected version of Cryptolocker was released under the name 'Cryptowall'. It was also able to hide on the infected system by infecting 'svchost', a legitimate windows process. It also maintained persistence on the system by creating a run entry [9, 12].

In 2014, Cryptowall 1.0 improved the malware further by communicating with a command-and-control centre (C&C) via HTTP using the RC4 algorithm. This version, though, had a shortcoming in the DeleteFile() API used by the malware, which did not entirely delete the files from the victim's machine, hence allowing cyber professionals to assist victims in recovering the infected files on the disk. In late 2014, Cryptowall 2.0 fixed this flaw and was able to delete users' files effectively. Hackers improved the malware to create a unique bitcoin address to track and ensure each victim had paid the ransomware. Moreover, they were able to encrypt about 146 types of file extensions [9, 13]. In early 2015, Cryptowall 3.0 was detected and spread using exploits such as Magnitude and Fiesta. The malware could encrypt about 312 file extensions. The victims had to pay the ransom by accessing the personal page of the attackers, which was only accessible via the Onion Router (Tor) [9, 14]. In late 2015, Cryptowall 4.0 spread through spam job application emails and via the Angler exploit [15, 16].

The most malicious ransomware attacks in recent memory, Petya and Cryptowall, were distributed using phishing campaigns. A phishing campaign consists of spam emails which may ask for sensitive information or contain malicious attachments. The impact of a phishing campaign is reliant on its sophistication. For instance, 'Dark Basin', a hack for hire group, had targeted several individuals on six continents during an extensive phishing campaign. The hacker group meticulously crafted the emails specifically for the target victims, as testified by one of the victims who expressed his concern about getting several spam emails with links that looked like they were coming from his family and friends and contained conversations of interest [17, 18].

In March 2016, 'Petya' ransomware spread through phishing emails containing resumes with an executable file targeting Ukraine. The malware infected the master boot record (MBR) overwriting the windows bootloader and then restarting the system. Upon restart, the payload encrypted the Master File Table (MFT) and followed with a ransom message to notify the victims of what just happened [19]. 'Red Petya' improved on its predecessor by encrypting the MFT using a more effective encryption algorithm, Salsa20. In September 2016, 'Green Petya' corrected additional flaws in 'Red Petya' and used a combination of Petya and Mischa virus code. If the malware failed to encrypt the MFT, as seen in Petya, it would then encrypt individual files using the Mischa virus. In December 2016, 'Goldeneye' (new version of Petya) appeared to strike Ukraine and Russia before spreading to the rest of Europe and beyond. It first encrypted files using Mischa followed by encrypting the MFT. Fast forward to June 2017, a new variant, 'NotPetya' brought down several organisations around the globe in a few hours [5].

The 'Grandcrab' was the most widespread ransomware of 2018, infecting more than 50,000 systems in a short period of time. Similar to 'Cerber', it used the ransomware as a service model (RaaS), allowing developers and hackers to share their ill-gotten profit. Grandcrab exploited Microsoft Office Macros and Powershell scripts to compromise the system. The operators of Grandcrab collected approximately $2 billion in ransom payments. In 2018, 'Ryuk' targeted big enterprises using military algorithms and collecting ransom in bitcoins. The use of cryptocurrencies gave an essential advantage to the attackers, as they made much easier for them to hide and remain anonymous [20, 21]. In 2019, 'Sodinokibi' (also referred to as Revil or Sodin) appeared in a web attack on the Italian WinRAR tool. It seemed to be a successor of the cyber espionage 'FruityArmor' (active since 2016) and had since massively impacted several countries including Taiwan, Germany, Italy and Spain. It also uses the RaaS model similar to Cerber and Grandcrab [21].

8.4 IMPACT ON ORGANISATION'S DURING COVID-19 PANDEMIC

Since the pandemic, several organisations worldwide have witnessed a massive surge of ransomware attacks. According to Check Point Research in Q3 2020, a 50% daily rise in ransomware attacks was observed compared to the first quarter of 2020 [22]. The ransomware attacks targeted several North American industries such as the government (15.4%), manufacturing (13.9%), construction (13.2%), utilities (11.1%), services (10.4%), retail (7.5%), hospitality (6.1%), healthcare (5.7%), education (5%) and financial (4.6%) [23].

In January 2020, Communication and Power Industries (CPI), a major electronics manufacturer for the defense and communications market, was hit by a ransomware attack. The domain admin, the highest privileges user in the company, accidentally clicked on a malicious link in the email, which infected the network. About 150 computers were still operating with Windows XP, which stopped receiving patches from Microsoft since 2014. CPI paid a ransom of about $500,000 to the attackers and obtained the key to decrypt a few servers [24, 25].

According to the '2020 Cyberthreat Defense' report, the number of organisations affected by the ransomware was about 62.4% compared to 2019 (56.1%) and 2018 (55.1%). In the last 12 months, about 57.5% organisations paid the ransom and only 66.9% received the decryption key to recover the data. Although cybercriminals are using innovative and more sophisticated methods to maintain anonymity and persistence to the system, companies worldwide have enhanced their cybersecurity by hiring cyber professionals to build a more robust defense mechanism against such cyber-attacks. About 84.5% organisation were able to recover the data without paying the ransomware in the last 12 months [26]. It is important to stress the fact that cybersecurity professionals advise against paying any ransom, as this may encourage the cybercriminals to conduct more of such attacks. Nevertheless, some companies may decide to accept to pay the ransom due to the sensitive nature of the data encrypted and/or stolen by the attackers, such as the University of California San Francisco (UCSF). In June 2020, in fact, UCSF was a victim of a ransomware attack, where the malware encrypted relevant academic work used to serve the public good. They paid approximately $1.14 million to recover the encrypted data [27].

The ransomware attacks may also cause reputational damage and financial damage to an organisation. In April 2020, Cognizant was infected with the Maze ransomware attack, asking for a large ransom. The company has spent about $3 million on recovering from damages and estimates to invest $50–$70 million to strengthen their cybersecurity [28]. The Maze cybercriminals are known to download the victim's servers' data and blackmail them into paying the ransom. The 'DoppelPaymer' hacker group uses a similar method. In November 2020, a major Taiwanese electronics manufacturer, Foxconn, was struck with a ransomware attack conducted by the latter criminals, which encrypted 1200 servers and stole 100 GB files plus 30 TB of backup. They demanded about $34.7 million in ransom. The DoppelPaymer cybercriminals warned the organisation that they would publicly release the encrypted data upon failure to pay the ransom [29].

These cybercriminals are known to follow a standard structure of attack called 'cyber kill chain'. In April 2020, the 'Magellan' phishing ransomware attack was a notable example of this approach [30].

The attackers first gathered information about the target, in what is called the 'reconnaissance' phase. During the following 'weaponisation' phase, the attackers would create the malware based on the vulnerabilities identified. The hackers would then use a sophisticated social engineering attack to access the targeted severs, known as the 'delivery and exploitation' phase. Once this was accomplished, the attackers had gained access to sensitive data such as employee ID numbers, social security numbers and more. At this point, they were able to install their malware to steal the Magellan employees' login passwords and finally launch the ransomware attack, in what is called the 'installation' phase. The following 'command-and-control' phase involves the infected machines communicating with the attacker's server. Finally, the last stage of the cyber kill chain is the attack's objectives, which was financial gain. However, in this specific case, the company identified the breach post April and employed a cybersecurity forensics firm, Mandian, to investigate the incident. In August, they reported a data breach of 1.7 million records, including internal staff and external customers [6].

Large enterprises have the resources to invest in updating and rebuilding their cybersecurity programmes. On the contrary, small and medium enterprises (SME's) often lack such resources to recover from ransomware attacks. With a massive shift to remote working, the secure corporate perimeter was broken, leaving more systems vulnerable to attacks, such as home networks and smart devices. While working from home, in fact, a ransomware attack affects both personal and working capabilities, leading to a more substantial downtime and possibly, substantial financial loss. The ransomware attack on Baltimore public school, for example, also affected their working from home capabilities and halted online sessions for more than 115,000 students [31].

During 2020, we clearly saw how attackers are disrupting remote working capabilities and conducting well-crafted phishing campaigns about COVID-19. Cybercriminals are developing malicious websites selling essentials such as hand sanitary, toilet papers and medicines that may be in short supplies. At the same time, cybercriminals such as the 'Revil' group auctioned stolen organisations data on the darknet. It also became more common for low-skilled attackers to work with other developers and create more sophisticated attacks and then split the profit among themselves, pioneering a model called 'ransomware as a service' (RaaS). Hackers may also modify existing malware or buy exploit kits to create more targeted cyber-attacks.

It should also be pointed out that attackers may not necessarily target a specific organisation but, rather, focus on a specific vulnerability to exploit a certain number of systems. As we discussed, when malware infects several machines, it can be controlled via the use of a command-and-control centre, forming a botnet. In 2016, the 'Locky' ransomware attack used the Necurs botnet to conduct a phishing campaign containing malicious email attachments. The email attachments included Powershell VB scripts, malicious Javascript and Microsoft Office VBA macros [32].

Often attackers also attempt to exploit the supply chain of an organisation. For instance, if an organisation A hires its SOC team from company B and its developers from company C, company B and C may also store Company A's information on their servers. A scenario where Company B or C are impacted by a data breach or ransomware attack can easily impact also company A. For instance, the 'Sunburst' malware compromised the SolwarWind's Orion software, which affected about 18,000 customers, impacting several organisations worldwide [33].

8.5 RISK MITIGATION

As discussed, ransomware attacks have been the most prevailing cyber threat for organisations worldwide during the pandemic by far.

To mitigate risks and fight ransomware attacks, companies are adopting different prevention and detection techniques. Prevention techniques could indeed help organisations overcome cyber-attacks and build more robust systems, while detection techniques can help organisations fight more effectively against them. Most importantly, though, there is a need to promote a constant effort to tackle the problem at its root. This means software engineers should have security in mind from the beginning of the development cycle and write more secure and easy to maintain software to begin with, while also constantly patching existing releases carefully and regularly.

Monitoring networks with efficient machine learning tools to spot intruders is also becoming more common and effective [34]. Training and simulations are also going to play an increasingly important role, as demonstrated by the US government who has even dedicated the whole Plum island, a small island off New York, to simulate cyber-attacks on its power grid and experimenting with ways for a quick recovery [35].

We have then two main areas where companies of any size must get prepared: prevention and then, detection and recovery.

8.5.1 Prevention

The most well-known prevention techniques are as follows:

- **Backups:** Organisations must take both online and offline backups. Online backups can be connected to the network or stored on a reliable cloud service. The offline backups stored on a server should not be connected to the network and entirely offline. For instance, shipping company Maersk, a NotPetya victim, was able to restart its services thanks to a backup on a server in Ghana, which remained offline due to power outage during the cyber-attack.

- **Patch systems:** The company own IT professional staff must regularly patch the organisation's software and operating system to overcome any disclosed vulnerability.

- **Identity and access control:** The organisation should grant access to resources based on the employee's role or requirement. Since employees work from home during the pandemic, defining resources' access would limit the damage during a cyber incident.

- **Multifactor authentication:** Several cyber professionals have emphasised that companies must adopt strong password policies. They must implement two-factor authentication, including using two identification methods, such as password and OTP authenticator, password and biometric authentication, etc. The use of multifactor authentication (2MFA) creates an additional layer of security. On the other hand, it has been demonstrated how forcing an overly articulated password creation policy may be counterproductive as users may then recourse to potentially unsafe workarounds to store and remember said passwords [36].

- **Separation and monitoring the network:** Since the pandemic, many professionals are working from home and are operating on the home network, it is vital to utilise a secure VPN while accessing the company's resources. The organisation must also ensure that critical assets are separated from the network, for example by using a so-called demilitarised zone (DMZ). If these precautions are adopted, during an incident a company may be able to secure and retain some of its crucial assets.

- **Awareness among employees:** The organisation's employees play a vital role to define an effective strategy for security as they are constantly targeted by ongoing phishing campaigns. Attackers, in fact, often target company professionals with spam emails that, if acted upon, may allow them to gain privileged access to the organisation's sensitive data. The employees must be able to distinguish a phishing email. The incoming emails must be filtered out as external and internal to detect phishing emails more efficiently. Relevant training sessions should be provided to all employees to let them learn about cyber risks and dangers. Red and blue teaming exercises could also help organisations manage and map ways to counter cyber incidents or new-fangled behaviour.

8.5.2 Detection and Recovery

The most well-known detection techniques are as follows:

- **Monitoring the network:** The organisation must control the incoming and outgoing traffic from the network utilising firewall rules. Integrating an intrusion detection system (IDS) and intrusion prevention system (IPS) would allow organisations to detect and filter out unknown anomalies instantaneously. The organisation must also use antivirus software which can alert if an intrusion occurs.

- **Red teaming and pen-testing:** The organisation must employ cyber professionals to perform external and internal penetration testing and red team exercises. The exercise would help organisations examine their network and systems' robustness so as to tighten it further and mitigate future attacks.

- **Recovery:** The organisation could form business continuity plans (BCP) and disaster recovery plans (DCP) for its systems and network to continue its services after a cyber-attack. Organisations could also hire cyber-incident responders to assist in the aftermath following a cyber-attack.

Usually, during a cyber-attack, organisations are recommended to disconnect all computer systems to avoid malware from spreading through the network. The first step is then to understand and determine the scope of infection, which should be followed by restoring the files from a backup. During a ransomware attack, the organisation can either decide to pay the ransom or not. In a scenario where the organisations choose to pay the ransom, it is also possible to hire ransomware negotiators, to negotiate the ransom amount with the attacker. Nonetheless, it should be reminded that the cybersecurity community at large advises against paying any ransom, to discourage attackers from performing such attacks again in future.

8.6 CONCLUSION

The future of ransomware attacks is likely in increasingly more sophisticated strategies for distributing malware and maintaining anonymity. Intruders can maintain anonymity in different ways, for example by presenting false flags to mislead the source of the attack. The attackers would then adopt several techniques to maintain persistence in the victim's machine and avoid detection. Attackers may also target an organisation based on the victim's demographics and/or specific assets. Therefore, it is essential to secure the company's supply chain in its entirety to protect its critical assets. In conclusion, going forward, we can expect cyber-attacks to be prevalent during a pandemic or any sort of international crises. Cybersecurity professionals have managed to tackle most dangers over the years and continue to provide organisations with a secure working environment. Remaining up-to-date with the latest news, discovered vulnerabilities and strategies is paramount for a successful and effective defence. From one cybersecurity professional to another, "we can protect organizations by helping each other."

NOTES

1. A hacker group that appeared in 2016. See https://en.wikipedia.org/wiki/The_Shadow_Brokers
2. This continues a long tradition of informing infected users that traces back to the 1980s, with examples like the SCA virus on the Commodore Amiga computer, which told its victims the much less dangerous but even more creepy message "Something wonderful has happened: your AMIGA is alive!!!"

REFERENCES

1. K. Mitnick, &. W. L. Simon, Ghost in the Wires: My Adventures as the World's Most Wanted Hacker, New York: Little, Brown and Company, 2011.
2. D. Bisson, "The TalkTalk Breach: Timeline of a Hack," State of Security, 2015.
3. R. D. Vere, "TalkTalk, one year later. The AntiSocial Engineer Limited," AntiSocial Engineer Limited, 11 2016.

4. K. Zetter, Countdown to Zero Day: Stuxnet and the Launch of the World's First Digital Weapon, Broadway Books, 2015.

5. A. Greenberg, *The Untold Story of NotPetya, the Most Devastating Cyberattack in History*, 2020.

6. C. F. &. Security, "WannaCry ransomware attacks cost the NHS £92m," *GlobalSign GMO Internet*, Inc, vol. 2018, pp. 1–3, 10, 2020.

7. G. Smith, "The original anti-piracy hack," *Retrieved Apr*, vol. 10, p. 2005, 2002.

8. N. Hampton and Z. A. Baig, "Ransomware: Emergence of the cyber-extortion menace," 2015.

9. A. Mohanta, K. Velmurugan and M. Hahad, Preventing Ransomware: Understand, Prevent, and Remediate Ransomware Attacks, Packt Publishing Ltd, 2018.

10. B. Donohue, *Reveton Ransomware Adds Password Purloining Function*, 2013.

11. ESET, "Virlock: First Self-Reproducing Ransomware is also a Shape Shifter,"WeLiveSecurity, 1 2014.

12. Y. Klijnsma, CryptoWall tracker: CryptoWall 1, 2019.

13. Y. Klijnsma, CryptoWall tracker:CryptoWall 2.0, 2019.

14. Y. Klijnsma, CryptoWall tracker: CryptoWall 3.0, 2019.

15. Y. Klijnsma, CryptoWall tracker: CryptoWall 4, 2019.

16. C. Brook, "Angler Exploit Kit Spreading Cryptowall 4.0 Ransomware," Threatpost, 12 2015.

17. J. Scott-Railton, A. Hulcoop, B.A. Razzak, B. Marczak, S. Anstis and R. Deibert, "Dark Basin: Uncovering a Massive Hack-For-Hire Operation," *Citizen Lab*, 8 2020.

18. U. S. Department of Justice, "Private Investigators Indicted In E-Mail Hacking Scheme," U.S. Attorney's Office Northern District of California, 2015.

19. S. Y. A. Fayi, "What Petya/NotPetya ransomware is and what its remidiations are," in Information Technology-New Generations, Springer, 2018, pp. 93–100.

20. E. U. A. f. Cybersecurity, "ENISA Threat Landscape Report 2018," European Union Agency for Cybersecurity, 2019.

21. E. U. A. f. Cybersecurity, "ENISA Threat Landscape 2020 – Ransomware," 2020.

22. Check Point Software Technologies LTD, "Global Surges in Ransomware Attacks – Check Point Software," Check Point Software, 10 2020.

23. R. D. Adams, "Infographic: Ransomware attacks by industry, continent, and more," TechRepublic, 2020.

24. A. Choudhury, "Top 8 Ransomware Attacks of 2020 That Shook The Internet," *Analytics India Magazine*, 12 2020.

25. Z. Whittaker, "Defense contractor CPI knocked offline by ransomware attack," *TechCrunch*, 3 2020.

26. L. CyberEdge Group, "CyberEdge 2020 Cyberthreat Defense Report CyberEdge Group," CyberEdge Group, LLC, 2020.

27. University of California San Francisco, *Update on IT Security Incident at UCSF*, UC San Francisco, 2020.

28. O. Johnson, "Cognizant Contains Maze Ransomware Attack as Cleanup Costs Spiral," *CRN*, 5 2020.

29. M. Novinson, "Foxconn Ransomware Attack Reportedly Damages Servers, Backups," *CRN*, 12 2020.

30. L. Spitzner, "Applying Security Awareness to the Cyber Kill Chain," *SANS Security Awareness*, 2 2019.

31. B. B. C. News, "Ransomware halts classes for 115,000 Baltimore pupils," *BBC News*, 11 2020.

32. M. Labs, "Ransom.Locky," 2020.

33. E. Gately, "Sophos: Impact of Massive SolarWinds Software Hack Still Unfolding," *Channel Futures*, 12 2020.

34. A. Greenberg, Sandworm: A New Era of Cyberwar and the Hunt for the Kremlin's Most Dangerous Hackers, Anchor, 2019.

35. C. Beatrice, "Photos: Plum Island, restricted US site where DARPA preps for cyber war - Business Insider," *Business Insider*, 2019.

36. R. Dillon, S. Chawla, D. Hristova, B. Goebl, S. Jovicic, "Password Policies vs. Usability: when do users go 'bananas'?", Proceedings of the IEEE 19th International Conference on Trust, Security and Privacy in Computing and Communications (TrustCom), pp.148–153, Guangzhou (China), 2020

The Effects and Innovations in Gaming and Digital Entertainment Forced by the Pandemic

Marco Accordi Rickards[1], Micaela Romanini[2] and Guglielmo De Gregori[2]

[1]*VIGAMUS Foundation; University of Rome Tor Vergata, Rome, Italy*
[2]*VIGAMUS Foundation, Rome, Italy*

CONTENTS

DOI: 10.1201/9781003148715-9

9.1 INTRODUCTION

With over 2.5 billion players worldwide, gaming has shown its incredible strength in bringing people together during the lockdown and the period of social distancing. If we think about multiplayer and online game modes, by definition virtual experiences, it is immediately evident how, on the occasion of the forced isolation due to COVID-19, this played a fundamental role in promoting social distancing. Even the World Health Organization has declared video games as a useful pastime in times of quarantine, supporting the #PlayApartTogether campaign, launched by numerous international publishers, such as Activision Blizzard, Amazon Appstore, Microsoft, Riot Games, Twitch, Ubisoft, Unity, Wooga, YouTube Gaming, Sega, Sony, Zynga, who took part in several fundraising programmes.

Thanks to the "Play at Home" campaign, Sony Playstation has made it possible to download for free some of its most popular sagas, such as Uncharted and the indie title Journey, much appreciated for its atmosphere. Microsoft too, with Xbox, through the "Support a hero, stay at home", rewarded the time spent on the console, fully applying gamification to the epidemic containment measures adopted by the various countries, transforming the objectives and rewards obtained by the players into donations in favour of subjects at the forefront of management of the COVID-19 emergency.

Nintendo and Razer have actively participated as donors, providing face masks and other personal protective equipment, lung ventilators and medical supplies but numerous brands have chosen to donate, including Ubisoft, Rockstar and hundreds of thousands of dollars donated to hospitals and other parties directly involved in the management of the pandemic.

9.2 GAMES AS THE ULTIMATE ENTERTAINMENT PLATFORM DURING LOCKDOWN

Foundraising was, moreover, a leitmotif of many initiatives for #playathome and #stayathome campaigns and, more generally, a hot topic in the management of the coronavirus emergency. Just from the gaming world, and not always strictly linked to the initiatives of the big players in the sector, original solutions have arrived such as Gamindo, an app available on the main marketplaces that allowed you to donate by playing: every game completed on the application was automatically transformed into a donation to intensive care units and COVID departments of the most affected hospitals. During the weeks of lockdown, the app has seen the multiplication not only of active users but also and above all the time of use.

Among the many trends that the health emergency linked to COVID-19 has brought about, the increase in sales volumes of the videogame sector and gaming-related products appears interesting: in the period March, April and May 2020 the sales of video games in the the United States has largely exceeded the turnover of previous years, marking a quarter dedicated to gaming. Suffice it to say that, if in May 2019 a turnover of 640 million dollars was achieved, in May 2020 it was close to 1 billion, totalling 980 million dollars.

In the five European countries most affected by the coronavirus emergency – Italy, Spain, France, Germany and the United Kingdom, visits and traffic on sites and apps dedicated to gaming increased on average by 19% in just one week, from 6–12 April.

In the same period, the sale of video games has marked record figures: in the first three days of its launch, Ghost of Tsushima by Sucker Punch has sold 2.4 million copies, The Last of Us II by Naughty Dog 4 million copies, Nintendo's Animal Crossing: New Horizons 11 million digital copies in 11 days, while Activision's Call of Duty: Warzone, the free chapter of the online saga since March 10, reached 60 million players in May alone. Among the best performing companies in the months of lockdown, Activision Blizzard is confirmed, which includes the Call of Duty galaxy but also the smartphone game Candy Crush; Peak Games, publisher of Toy Blast, Township and Wildscapes; Supercell, brand that brings together Brawl Stars and Clash of Clans.

According to Superdata, the global digital video game market reached an all-time high of 10.54 billion dollars revenue in April 2020, before falling to 10.46 billion in June, compared to 9.6 billion dollars reached instead in June 2020.

Between April, the time of greatest traffic growth, and June 2020, Activison records a drop of 9.4% in connected users and 13.6% of the average stay time per visitor (calculated on a monthly basis). Peak Games users drop by 21.3% and the average number of minutes spent is down by 8.2%. Supercell, on the other hand, is growing by 4.2% of new gamers but cannot compete with the summer and the average play lasts 23.7% less.

Video game streaming sites also reached a large number of visits and views during the social distancing period: in the first quarter of 2020, Twitch exceeded the overall three billion hours viewed for the first time. Other data, reported by the Times, speak of a dizzying growth in the sale of hardware and accessories for gaming, which in March alone would have exceeded one and half billion in values.

Research also highlight extremely encouraging data linked to the development of digitisation in several countries, such as Italy: in a few weeks and in some cases in a few days, the massive use of technology in the daily practice of companies, institutions and individuals has allowed a turn towards digital. There has therefore been an increase in internet users, in the intensity of the data flow and in household equipment, for professional needs or "safe" purchases but also for gaming and entertainment on demand.

Online gaming has established itself as the most loved hobby during the lockdown but this trend is destined to continue: once the lockdown experience is over, people will have transformed their homes in the centre of their daily life, also following the reopening of restaurants and outdoor activities. These data indicate a growth certainly accelerated by the pandemic but also a new trend in use, a change in the behaviour of individuals. With the release of platforms such as Google Stadia, the consolidation of PlayStationNow and Xbox Game Pass, the migration towards a future now dominated by cloud and digital platforms has been confirmed.

If the game resides on the cloud platform, there are no longer device limitations: the new scenario is one of interaction between one object and another, as for music or editorial consumption, thanks to the advent of new generation smartphones, characterised by an ever greater computing power, which allow a diversified use.

According to media expert Matthew Ball, the explosion of gaming during the lockdown period has simply accelerated a process already underway, which sees gaming as the most successful media of this historical period. According to Ball, the reasons for success lie in various factors: of great importance, the economic one, for which video games, now interoperable, does not necessarily require the purchase of ad hoc devices. A considerable advantage in a time of economic difficulty like the one we are still experiencing. Secondly, the narrative universes of video games are more versatile and more easily adaptable to the present, with the situation we have experienced, which is very reminiscent of sagas with a post-apocalyptic theme. Ball also shows optimism towards the production of video games, characterised by a great variety of small and medium-sized enterprises, unlike other entertainment industries: even if with a little difficulty, small independent producers should be able to survive the post-economic crisis.

9.3 A SHIFT OF PARADIGM IN GAME AND TECHNOLOGY COMPANIES

Among the most remarkable effects of the recent pandemic on the global video game market, there's a change in the way video games and softwares themselves are created, namely the remote, work-from-home model, adopted by many studios in the world, proved itself to be very effective when it came to video game creation. This very smooth transition shouldn't be of any surprise, as game developers have historically been used to work from their houses, especially when it comes to indie productions and "garage games". International collaborations, supported by remote project management and version control systems, such as Trello, Slack, Discord and GitHub that were already largely in use between game developers even in the pre-pandemic world, as this allowed smaller studios, especially in Countries where the games industry is not as developed as in Europe or North America to get in touch with other developers and freelancers from all over the world and work on game titles without the need to be physically present. On the other hand, the game developers landscape has been always inevitably connected to a more creative and freedom oriented attitude, as game creation isn't necessarily connected to traditional work dynamics observable in other industries; last but not least, creators and developers were already very familiar with not just the classic online collaboration tools but even with more advanced solutions, which involved virtual and mixed reality. Ironically, game developers who were among the first unwitting prophets of pandemic and global emergency scenario, spectacularly depicted in games such as Fallout and The Last of Us or in books like Ready Player One, proved to be also the quickest and most adaptable subjects when the necessity arise to shift from a "in real life" routine to a much more secluded environment. Some people, amidst the extremely unfortunate occurrence they ended up living in, on the other hand found ways to find new ways to communicate with their colleagues and stay mentally healthy while overcoming the necessary evil of the social distancing. While new platforms like Zoom showed up providentially during the pandemic (mainly because of the simpleness of their setup) and services like Google Meet were largely improved in their functionality, many game developers and members of the software and technology industry used the already existing virtual reality (VR) technology to ease the disconnect that derives from a "screen only" setup. After the idea of setting up a meeting inside an online environment has been now widely recognised as a staple of a workplace in the post-pandemic era, one of the most natural evolution of older video-based platforms is a new kind of platform which heavily rely on VR and similar system which are able to introduce empathy in online social interactions. While since many years as of now VR platforms have stopped being just a weird novelty, as the interest (and the research initiatives) in this field have been strongly reignited after the reinvention of the medium operated by Oculus in 2012, virtual meeting technology were probably a too much difficult setup to create for a common office, especially outside of the tech industry. However, many companies, such as the VR startup Spatial[1] have spotted the opportunity of creating a platform which is able to guarantee not just a voice or video presence but a real simulation of physical presence, something that it's not too much different from what we have seen in movies such as Star Wars and was considered until a few years ago a mere science fiction gadget. The main catch of a VR meeting resides in the possibility for people to express through body language and facial expressions, something that while is still certainly possible in video calls, yet feels very unrealistic and soulless, compared to the possibility of donning a VR helmet and see the other person in a three dimensional view.

Just like the Star Trek PADDs eventually became the blueprint for Apple's iPad, it's not too farfetched to affirm that offices in an interconnected world could look much more similar to a Star Wars movie than we expected and definitely much less physical. Of course, we didn't need to wait for a grim scenario like the COVID-19 pandemic to drive and accelerate such technology. On the contrary, just like Palmer Luckey from Oculus recognised back in 2012, VR is a tool that can improve the lives of millions of people in the world, especially a hyperconnected one. As global warming proved to be a possible threat in the upcoming decades, thus calling the global population to a more responsible use of airplane flights and a more conscious observation of emissions deriving from fossil fuel, VR could be a way to neatly escape the necessity for many workers to travel so frequently for their business meetings or conferences. Conversely, as the global pandemic has changed the habits of people, introducing concepts such as "staycation" (the idea of working from remote, leisure locations) and causing a dramatic rise of new non-physical based workers in the job market (e.g., streamers, influencers and VTubers), it's safe to assume that people are feeling more keen to travel than ever and willing to behave as citizens of the world, especially after being secluded for so many months; companies, on the other hand, are starting to realise that many of their people are more than ever willing to change their physical seat many times in the course of their life, thus requiring a new attention from human resources when it comes to the allocation of workers among the different global branches of a company, thus creating the need of smooth communication tools that should allow a game development studio based, say, in Hong Kong, to keep a smooth production pipeline by working on a daily basis with an office in Chennai. To accommodate all of this needs, contextually arising in the process of shifting to a more nomadic, sustainable model of work–life balance, VR and similar devices could be a way to create a "portable office environment" which can be accessed anywhere in the world.

9.4 UNREAL ENGINE AND THE NEW WAVE OF FILM AND GAMES PRODUCTION

The Future Forum research observed how the majority of knowledge workers inside a sample of 4,700 subjects would never want to go back to the old way of working. Only 12% want to return to full-time office work and 72% want a hybrid remote-office model moving forward[2]. This should be largely be attributed to the massive shift in the global perception of life, which is causing people to make new choices about where and how they want to live, thus creating new expectations and higher standard when it comes to flexibility, working conditions and work–life balance. Even in this case, game studios are already posed on a fast track when it comes to adapt their routines to the "new normal". The huge success of The Mandalorian, the most recent Star Wars spin-off produced by Disney and streamed on the Disney+ platform is certainly to be attributed also to the adoption of Unreal Engine and virtual sets for the creation of the series and its most iconic moments. Unreal Engine, the highly performant mod creation software tool by Epic turned professional game engine, is certainly one of the most widely adopted solution for game studios of any size all over the world; however, this game engine was largely used by The Mandalorian's creative minds in order to iterate fast on the scenes, while actors moved in front of giant LCD screens where the game engine almost photorealistic output was displayed, all in front of traditional movie cameras. This innovative pipeline provided the opportunity for the director and its crew to orchestrate the sequences many times and find the best tone and angle, without the need of reshaping the set at every single shot. While luckily The Mandalorian's season two production wasn't affected by the pandemic, many observed how the Unreal Engine pipeline could help the renaissance of the movie industry while the world adjusts to the post-pandemic effects[3] with physical distancing still a need in many parts of the world and of course the significant cut of budgets that the temporary closure of theatres has caused on movie industry as a whole. While the creation of 3D environments and the large use of green screen inside movies is certainly not something new, the solution offered by Unreal Engine and pioneered by Disney and ILM has shown a wider potential since it eschews the traditional post-production pipeline, which necessarily always created a temporal disconnect between the shooting and the compositing of actors on the 3D environment; in other words, we can now assist at the rise of new professional figures, such as the "virtual set designer", someone who's capable to fastly iterate movie sequences with the user-friendly approach of a game creation engine. As games and movie productions tools, pipelines and best practices are closer than ever, we've ushered into a further revolution of the entertainment industry, where the borders between the different medium are more blurred than ever. Certainly, the pandemic has shown us a world which is much less physical than we remembered before, as the dramatic spike in video game sales has largely demonstrated, and certainly the collective perception was shifted in a direction that eventually reinforced and consolidated the idea of the online, digital and non-physical world as something that is not a mere mirror of physical reality, but instead a widely interconnected ecosystem which can be considered a sole entity together with the real world. The ability of creating a virtual world in software like Unreal Engine and deliver it to a completely different part of the globe, where it will be used to shoot a new pilot of a TV series, an immersive VR video game or even a phyrtual theme park experience, is one of the many results of an year-long shifting process which ultimately led to a hyperconnected global working environment, whose dramatic rise was certainly accelerated by the COVID-19 crisis, though at the same time it was inevitably bound to change the world since the mid-2000s, at a time where the increasingly massive adoption of Internet change mindsets, point of views and lifestyles in every part of the world.

NOTES

1. https://www.voices.com/blog/virtual-reality-pandemic/
2. https://slack.com/intl/it-it/blog/collaboration/workplace-transformation-in-the-wake-of-covid-19
3. https://www.cnbc.com/2020/09/03/coronavirus-mandalorian-tech-key-to-jump-start-film-production.html

BIBLIOGRAPHY

1. Newzoo, Games Trends to Watch in 2021, https://newzoo.com/insights/articles/newzoos-games-trends-to-watch-in-2021
2. Newzoo, Global Games Market Report, 2020 https://newzoo.com/products/reports/global-games-market-report/

3. Newzoo, Global eSports Market Report, 2020 https://newzoo.com/products/reports/global-esports-market-report/

4. Newzoo, Global Mobile Market Report, 2020 https://newzoo.com/products/reports/global-mobile-market-report/

5. Businesswire, Games Industry Unites to Promote World Health Organization Messages Against COVID-19; Launch #PlayApartTogether Campaign https://www.businesswire.com/news/home/20200328005018/en/Games-Industry-Unites-to-Promote-World-Health-Organization-Messages-Against-COVID-19-Launch-PlayApartTogether-Campaign.

6. Superdata, Worldwide digital games market: December 2020 https://www.superdataresearch.com/blog/worldwide-digital-games-market

7. Technology used to film 'The Mandalorian' could ease Covid-19 film production woes https://www.cnbc.com/2020/09/03/coronavirus-mandalorian-tech-key-to-jump-start-film-production.html

8. Moving beyond remote: Workplace transformation in the wake of Covid-19 https://slack.com/intl/it-it/blog/collaboration/workplace-transformation-in-the-wake-of-covid-19

9. Virtual Reality in 2020: How Life Went Virtual During the Pandemic https://www.voices.com/blog/virtual-reality-pandemic/

III

Planet

While global human activities have been disrupted due to COVID-19, some would posit that such disruption has provided new opportunities for the planet. It is time to seize the moment and rethink alternative strategies towards the ways activities are organised and conducted, for example on education, city planning and workplace transformation.

CHAPTER 10

Digital Workplace Adaptation and the 20-Minute City

How Design Can Help Redefine the Future of Cities in a Post-COVID World

David J. Calkins

Gensler, Singapore

CONTENTS

DOI: 10.1201/9781003148715-10

10.1 INTRODUCTION

Starting in 2020, the COVID-19 pandemic impacted the way of life for most if not all of the world's population. Loss and disruption could be seen and felt everywhere as the pandemic spread to every country on earth. Yet, even amid the negative and costly experience, architects and designers found opportunities in the digital transformation that occurred through necessity as individuals, companies and governments struggled to find ways to remain connected and productive while attempting to protect human health. In the rapidly changing environment caused by the pandemic, designers have engaged in research and are working with clients to embrace change and help shape better urban environments that respond to new ways of working and living, that are technology-enabled and potentially healthier, more satisfying and more productive, offering elevated experiences, encouraging of greater diversity, and are more environmentally sustainable and resilient.

This chapter will examine the rapidly evolving nature of office work, including Gensler's digital transformation, COVID shaped attitudes of the people working in office environments and the potential transformation of office spaces, and its impact on post-COVID urban environments, including our strategies for developing more livable cities.

10.2 TRANSFORMATION OF WORK

The pandemic drove the biggest and most sudden workplace shift in history as millions of workers started working from home through the majority of 2020 and beyond, even though effective vaccinations were being administered in many countries in early 2021. Driven by the pandemic, workplace mobility and remote working moved from concept to universal implementation with barely a weekend to prepare.

The digital transformation of the work experience and the way business is done was accelerated tremendously by the pandemic with COVID acting like a time machine that accelerated change. According to the global consulting firm McKinsey (www.McKinsey.com), in just a few months, the COVID-19 crisis has brought about years of change in the way companies in all sectors and regions do business, particularly in the digital realm. In a McKinsey Global Survey of executives (https://www.mckinsey.com/business-functions/strategy-and-corporate-finance/our-insights/how-covid-19-has-pushed-companies-over-the-technology-tipping-point-and-transformed-business-forever), companies during the pandemic have accelerated the digitisation of their customer and of their internal operations by 3–4 years. The share of digital or digitally enabled products in their portfolios has been accelerated by 7 years. Survey respondents expect most of these changes to be long-lasting and are already making the kinds of investments that ensure they will continue.

Remote working and collaboration appear in the survey as the most frequently cited change being experienced by 93% of the respondents with at least 54% of the companies expecting that remote working will continue post pandemic. It's truly remarkable how quickly the shift to remote working happened, reportedly 43 times faster than expected with the new work methodology being implemented in about 10 days versus than the anticipated 454 days.

Given how quickly most companies changed to digitally technology-facilitated working from home, there was virtually no time to rehearse for full-scale virtual communication, collaboration and interaction, nor to test and set cultural and procedural norms, the impact to company culture and performance must be assessed and adjustments implemented. We are convinced that some form of remote working, including optional working from home will persist post pandemic. Given that, companies will have to decide what they want their hybrid physical/virtual cultures to be based on the big picture values of their organisations and then set appropriate HR policies to support them. For instance, when it is safe to do so, which staff members should come back to the office and how often? Certainly, physical space will also be rethought based on the need to support the hybrid work model.

Our firm's success in pivoting to full working-from-home mode was made possible by an advanced digital technology roadmap and significant strategic investment. Gensler's digital roadmap was implemented over several years to allow for infinite possibilities in growing our design practice, and it includes the development of a cloud-based ecosystem, the adoption of Building Information Modelling and computational design, the development of cutting edge visualisation and the performance-driven design solutions. The firm made significant capital investments in hardware and infrastructure, including providing laptop computers for every staff member. All of us have the ability, given reasonable internet access, to work anywhere in the world, accessing the Gensler network.

Gensler's advanced planning and investment allowed us over one weekend early in March of 2020 to close nearly all of our 50 global offices and to have our approximately 5,500 team members to work from home, collaborate and remain productive, while continuously working on our clients' project. Many other design firms were not as prepared and therefore struggled to adjust their ways of working. Our transition was seamless, as our Design Technology team rapidly migrated thousands of project files to the cloud as the final preparation, making them available to all. Since then, nearly our entire staff has continued work remotely and mainly in our homes, communicating internally and with clients on several collaboration platforms including Microsoft Teams, MS Whiteboard, Miro, Go to Meeting, and Zoom. All of our firm wide leadership meetings which occur several times each year have also been conducted virtually.

Through the pandemic, we have been continuously evolving the way digital technology assists our design process. In an era of fear that artificial intelligence will displace more and more jobs, our approach to utilising technology is not to replace humans in the design process but rather to enhance the uniquely human qualities our designers possess of insight, empathy and connection. We seek to endow our designers with superpowers of analysis and decision making through various new digital tools, including planning, simulation and visualisation. This combination of automation and human interaction is at the heart of our NFORM™ ecosystem, a platform that combines information metrics and design expertise. With this platform, we're now able to leverage Gensler's expertise in designing billions of square feet of workplace interiors, as well as decades of data gleaned from our about different types of spaces and how people use them, to help our clients navigate the future workplace.

Traditionally, design and modelling activities have been fundamentally disconnected from computational intelligence. Our technology ecosystem

FIGURE 10.1 Gensler gFloorz space planning and analysis tools customizable dashboard

has changed that equation by giving designers and clients real-time access to an expansive set of design variables so they can make informed and predictable decisions balancing form, function and business demands.

One such tool is gFloorz™. A key computational product in the NFORM ecosystem, gFloorz allows clients to leverage Gensler's history of data and combine that with their bespoke metrics to test fit interior

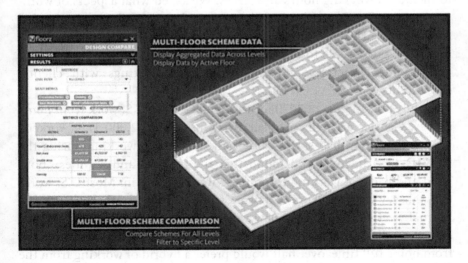

FIGURE 10.2 Example of gFloorz space plan and metric comparison display

workplace projects. Designers and clients can work together in real time to collaborate and explore options quickly, test assumptions, rapidly iterate spatial layouts, add or refine variables and understand the implications of design decisions early in the design process. We are able to help our clients to make rapid informed decisions to analyse, plan and design their workplace so they can seamlessly pivot and adapt to the changing nature of work.

10.3 SURVEYING ATTITUDES TOWARD WORK DURING COVID

As the pandemic has developed, we have studied the developing attitudes of our staff and those at other companies through targeted surveys. While attitudes seem to have shifted a bit over time, it seems clear that moving into the future, workers will expect to have the flexibility continue to work from home at least for some part of the week. Some people are opting for an entirely remote work engagement, allowing them to work in lower cost, less dense areas and evaluating the added benefits with potentially lower compensation.

A few weeks into the pandemic, we surveyed the staff in our Gensler Asia Pacific Middle-East Region and our results were surprisingly positive. Working from home, 68% of our staff was satisfied or very satisfied with the experience, 64% felt more productive, 67% feel more empowered and 55% felt more engaged. We also had negative feedback relating to increased working hours and stress. We asked what aspects of working in the office our team members missed most and 64% responded socialising with colleagues, 57% listed impromptu face-to-face meetings, 51% said access to ergonomic set-up for work, and 67% of Office and Studio Directors, scheduled in-person meetings with clients/consultants.

In July and August of 2020 for the 2020 Gensler U.S. work from home survey, Gensler anonymously surveyed more than 2,300 U.S. workers, who at that point had spent more than six months working during the COVID-19 pandemic, and found that the impacts and realities—both good and bad—of working from home were settling in. Most U.S. workers said they would still prefer to work from the office for the majority of a normal week but they stated new expectations around flexibility, privacy and space sharing as they return. Only 19% of U.S. workers want to work from home full time; over half would prefer a hybrid of working from the office and from their home as they look to the future.

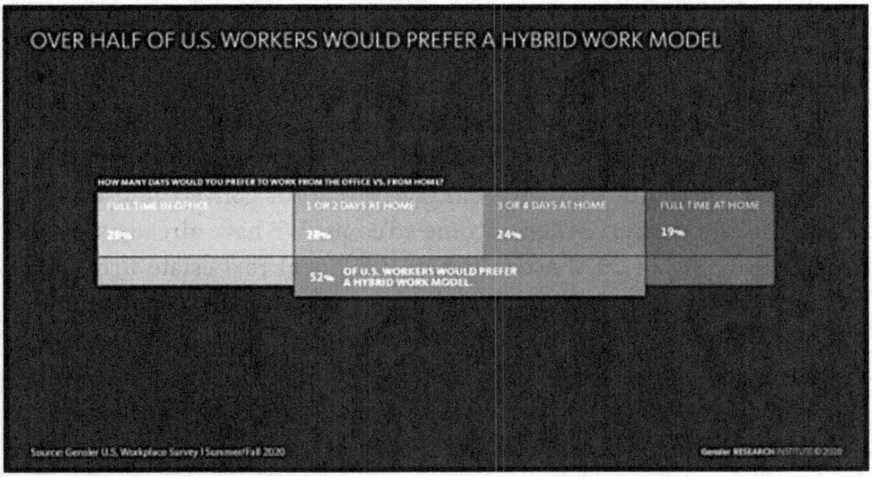

FIGURE 10.3   Responses to Hybrid work question – Gensler US Workplace Survey

What their responses seem to mean for the future of work and the workplace:

1. U.S. workers continue to see the physical workplace and in-person collaboration as key aspects of their jobs and driving reasons behind their desire to return.

2. They want to return to the workplace to collaborate and socialise, and for the long-term positive impact on their careers and organisational relationships.

3. Many of the benefits of working from home including greater flexibility and access to privacy in particular, will need to be reflected in the future workplace.

In summary, U.S. workers want to return to the workplace while keeping the benefits of flexibility and access to privacy they've enjoyed while working from home.

10.4 THE POST-COVID-19 OFFICE

The post-COVID-19 office may be significantly different based on a redefinition of its purpose. The office most likely will provide for the facilitation of collaboration and social interaction with significantly more technology embedded. One's cell phone will act as personal concierge providing greater personal control of the work experience. The office will also provide a greater meaning as brand expression and client hospitality hub. Companies may occupy less space as individual/focus work moves at least partially to homes or third places in the community, and this trend may lead to significant changes in downtown areas, and landlords convert unused space to other uses to preserve their income streams. We have already seen this trend developing in 2020 according to the global real estate firm CBRE (www.cbre.com). Through the first three quarters of the year, more retail and office space in the United States was vacated than occupied, leading to a negative absorption rate of those space types. This can be directly linked to the digital transformation of retail, which has seen a huge jump in online sales activity and the increase in working from home.

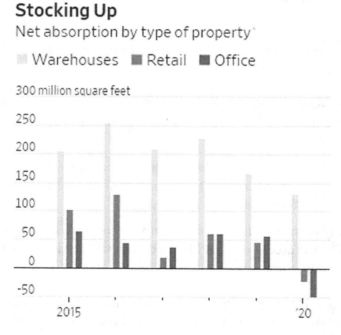

Stocking Up
Net absorption by type of property

■ Warehouses ■ Retail ■ Office

Newly leased minus newly vacated. Negative values means more space was vacated than newly occupied. 2020 is through third quarter.
Source: CBRE

FIGURE 10.4 Yearly lease space absorption by type

We now know that the overall effectiveness of virtual collaboration relies on a critical factor: human connection. Our recent work-from-home experiment has shown how important virtual collaboration platforms and management tools can be; however, nothing can replace face-to-face time with colleagues. Those interactions build social capital and personal connections that can keep us connected outside of the office.

As designers, we see the role of the office as the catalyst for engagement, inspiration and human connection and as a platform for meetings inspired by hospitality, collaboration and technology that fosters relationships and exchanges. Before the pandemic, awareness was growing around the concept of the hybrid workplace model which promotes collaboration, advanced technology, unassigned seating and activity-based design and offers a comfortable atmosphere driven by elements of hospitality. However, in the past, this model was limited by spatial conditions that required designing for a certain capacity and this, combined with a focus on space efficiency, resulted in increased densities. We have been reducing the amount of space per person/desk consistently over the past 30 years—but the global health crisis is rapidly changing this trend.

10.5 VIRTUAL EXPERIENCE AND CITIES

Our physical experience of cities has been restricted during the pandemic as we have spent most of 2020 relating to each other virtually through the use of digital technology. We have worked, met, shopped, studied and even celebrated with the family online. We have done these things for many months in the cause of keeping ourselves safe and healthy, and in that time we have probably formed many new habits, but our COVID-19 avoidance hasn't been the only force driving us to a virtual experience. It appears likely that in the post-COVID-19 pandemic future, our engagement with virtual experience will continue.

This is illustrated in a recent *Wall Street Journal* article titled "COVID-19 Propelled Businesses into the Future: Ready or Not" (https://www.wsj.com/articles/covid-19-propelled-businesses-into-the-future-ready-or-not-11608958806).

In many ways, digitization is simply the next chapter of a process underway for a century: the dematerialization of the economy. As agriculture gave way to manufacturing and then services, the share of economic value derived from tangible material and muscle shrunk while the share derived from information and brains

> grew … But the dominant driver is information technology. Joel Mokyr, an economic historian at Northwestern University, said one of its most important and least appreciated roles is the "great fake": It enables "increasingly accurate lifelike representations of some kind of reality through analog or digital mimicry, what you could call virtualization."

Digitisation and the utility and convenience that "great fakes" offer will continue to be formative, and will inevitably shape our lives; but we are not digital beings, we crave authentic in-person experiences. We want to be with other people. We long to take part in all of the activities that great cities have to offer in person. Designers are thinking about the opportunities that digital transformation is providing and requiring to shape our cities and redefine urban experience.

Cities are important to the world in very many ways. Today, roughly 55% of the world's population lives in cities. More than 80% of the world's GDP is generated in cities. They are hubs of business, culture, innovation and creativity. Cities serve as the most fundamental expression of density, and they will continue to attract all kinds of people because of their ability to enable community, connection and convenience. Cities will always be the engines of our Earth and need our highest level of attention because of their impact and potential.

10.6 MAKING MORE LIVEABLE CITIES POST-COVID-19

The pandemic has reminded us how important human health is for the economic health of the world. Because of their impact, healthy liveable cities are the key to increased prosperity and well-being in the world. Gensler has been thinking deeply about the ways design can help to positively redefine the future of cities in a post-COVID world by making them healthier, more equitable and happier places. In following sections we describe nine strategies we have developed.

10.6.1 Creating a Sustainable, Inclusive and Regenerative Urban Environment

This strategy involves redefining urban processes to be more like processes in nature by converting from a linear input/output system that takes in raw materials and puts out products and waste/pollution to a more circular/closed loop that reuses waste. The goal is for the city to no longer damages the environment but rather actively restores it, using smart technologies to maximise the efficiency of resource utilisation. We use design also is focused on human comfort, safety, walkability, flexibility, social opportunities and proximity to nature, nurturing the senses, creating delight and producing happy moments.

10.6.2 Embracing Economic and Generational Diversity to Create Inclusive Cities

This strategy expands opportunities and encourages affordable intergenerational living and connections. People from multiple generations and backgrounds coexisting makes for a richer urban experience.

10.6.3 Encouraging Polycentricity, Density and Compactness

Move from a single-use development mindset to a mixed-use development mindset to foster interaction, increase efficient use of infrastructure and enrich pedestrian journeys. This enables density, proximity, community connection, fosters creativity and affordability, and makes provision of community services and infrastructure viable. Technology has lifted the constraint of physical proximity by facilitating remote working and multiple urban centres reduce the need to commute and reduce the cost of living.

10.6.4 Leveraging Technology to Create Intelligent Cities

With people at the heart of the process, create a flexible Information Technology platform using data to inform decision making for cities. The desired result is a vibrant and flexible environment, linking people, processes and data, while respecting the cultural, social and natural characteristics of a place.

10.6.5 Responding to Changing Needs through Flexibility in the Public and Private Realms

Creating value and identity, increasing proximity to open space, improving sense of community, air quality and thermal control, understanding that every individual part is connected to a larger system of assets. Provide places for public art, emphasise health and wellness and provide for day and night use. Flexibility and adaptability in its DNA and taking back the streets as driverless cars become common. Treat streets as outdoor urban rooms to be inhabited by people.

10.6.6 Reimagining the Central Business District

Working from home is an opportunity to remake cities away from the industrial age model of separate zones. Central business districts (CBDs) can become more mixed, reducing commutes while making cities more affordable and inclusive. Major opportunities exist to increase urban permeability, variety in physical form, greenery, activity at the ground level, multi-level connectivity, human comfort and safety and security. A larger description follows.

10.6.7 Building 20-Minute Cities

Develop neighbourhoods that provide all of the necessary services to support daily life within a 15–20-minute walk from home. A larger description follows.

10.6.8 Investing in Public Transportation and Infrastructure

Leveraging public dollars with private sector investment to reimagine and expand transportation assets.

10.6.9 Fostering Public Partnerships for Urban Policy and Design

Increasing community participation when decisions are made to pursue equitable, inclusive and relevant design solutions.

Of these nine strategies, two potentially have the greatest impacts on the future of our cities. The following is our firm's thinking on **Reimagining the Central Business District** and **Building 20-Minute Neighbourhoods**. The next sections discuss these two in detail.

10.7 REIMAGINING THE CENTRAL BUSINESS DISTRICT

10.7.1 Impacts of the Pandemic on the Central Business District

Although CBDs are dense, connected areas that often serve as the centre point for regional transportation systems, most CBDs are imbalanced and weighted toward core commercial uses and business functions. As such, they are mainly populated on weekdays from 8:00 am to 6:00 pm, and as lockdowns and working from home became universal, impacts of COVID-19 on CBDs were particularly pronounced. As we head back to the office, a new hybrid style of work will emerge with many employees working from home for part of the week and returning to the office for the remainder of the week. Others might work predominantly remotely and come into office just a few times a month for specific tasks. This new hybrid coupled with greater interest in providing satellite office facilities located outside of CBDs, and the reduced patronage of supporting business like restaurants dependent on population in a particular area, will likely reduce the quantity of space occupied, reducing the activity and vitality of CBDs.

In many ways, the pandemic has accelerated trends that had already begun to transform our CBDs with a greater mix of uses: more focus on experience and destination, more responsiveness to evolving work patterns and mobility and the migration of families and companies from global/megacities to the suburbs, second-tier cities, satellite cities and less expensive metros.

10.7.2 The CBD Reimagined as Central District

The following are ideas for a reimagined central district not solely driven by office worker activity. These scenarios envision an optimistic future for our downtown cores.

More mixed-use and more local: Lower rents and property values can make cities more accessible and affordable to locals, with the opportunity to create more walkable "20-minute neighbourhoods." More residential and pedestrian-oriented uses with more green space can be infused into CBDs. Lower rents can also attract local startups, mom-and-pops, not-for-profits, small businesses, innovative food & beverage, community partnerships, makerspaces and local manufacturing. These homegrown enterprises exist in synergistic contrast to the reemergence of corporate workplaces in the CBD.

Hotbed for innovation: Empty spaces and lower rents will allow entrepreneurs to try new things in novel combinations. R&D and tech can also flourish, building off of innovation districts worldwide and their ties and proximity to higher education institutions. Creatively reimagining abandoned storefronts can attract culture, arts, diversity and quirkiness, seeding a new and authentic DNA. This will require a fundamental rethink of ground floor retail to allow for more public accessibility with integrated public/private and indoor/outdoor transitions. Consequently, a younger, more diverse population moves in. Streets become more pedestrian-oriented and lively.

Evolution of the employment core: As companies evolve to support a new model of working at the office, so must their neighbourhoods. Based on our recent research, one possible scenario is that as fewer workers travel daily, CBDs have an opportunity to become more appealing, hospitality-oriented and highly amenitised destinations that are more 24/7 and less 9–5. Parks and green spaces can flourish, promoting health and wellness. New kinds of assembly venues and entertainment options emerge.

How best for planners, designers, owners and developers to work together to achieve these future visions? Here are Gensler's eight strategies that can help create a more inclusive, resilient, sustainable, vibrant and healthy "Central District":

1. Rethink zoning and development regulations: To reshape the CBD, a change of mindset is critical to create zoning and development regulations that promote walkability, resiliency through adaptive reuse and inclusivity that would allow the CBD to be enjoyed by all.

2. Greater emphasis on public–private partnerships: With more public–private partnerships and an increased private sector role shaping a new CBD experience, companies and industries can help reposition cities and drive meaningful change.

3. Creative repositioning: Diversity in many aspects (in mixed uses, housing types, job creation, etc.) at many different levels (city, neighbourhood and building) is essential. As landlords adapt large commercial floor plates to accommodate different scales and uses, buildings are "hacked" to accommodate a new mix of functions. A similar dynamic is played out at the city block and CBD scales, further breaking down traditional typologies and zoning.

4. Prioritising diversity for greater resilience: We believe there is a strong correlation between urban regeneration and resilience. Urban regeneration is the process of renewal, restoration and growth that makes ecosystems resilient. And urban resilience is the capacity of people, communities, institutions, businesses and systems within a community to survive, adapt and thrive—no matter what kinds of chronic stresses and acute shocks they experience. The new CBD must diversify risk. In doing so, the future central district will be far more interesting, diverse, healthy, sustainable and fun.

5. Re-examining and rewriting regulatory and fiscal policies: To increase affordability and equity, support businesses to remain open

and stay, and foster job creation. All of these interventions would aim to attract and retain urban residents.

6. Investing in technology in its many forms (autonomous vehicles, new energy sources, new construction methods or new construction materials, machine learning, etc.) to increase safety, lower costs and lower our environmental impact.

7. Encouraging community participation to include all voices in shaping the future of CBD, and ultimately reshaping our future cities.

8. Open-mindedness to trying new things: In an age of profound uncertainty and change, everything has to be on the table. This will take a mindset of courage and optimism. While we think big, we must also prioritise smaller, less ambitious ideas that have an incremental but meaningful impact.

10.8 BUILDING 20-MINUTE NEIGHBOURHOODS

10.8.1 A Polycentric Planning Model

In addition to reimagining our CBDs as "central districts", implementing a polycentric planning model, sometimes known as the "20-minute city" can have very desirable impacts including efficiency, opportunities for wellness and community connection in our post-COVID-19 quest to make our cities healthier and more liveable. The 20-minute city has the potential to improve the quality of life, promote walking and free up space for other uses, such as parks and gardens.

This planning concept isn't new. It's really an early 20th century model of urban design in which residents' needs, from housing to school and work and leisure activities, can all be found within a 20-minute walk or short bike ride from their homes. In its new application, however, large cities are redeveloped into more self-sufficient districts, and the idea is extending to cities around the world, such as Barcelona, London, Paris and Singapore who are looking to implement a 20-minute city.

Some of the ongoing interest in technology-enabled working from home is being driven by health concerns over using mass transit systems, but people who have now lived for almost a year with no commute also see the time they have gained in lieu of commuting as a great lifestyle plus. This unwillingness to commute has revived the interest in 20-minute cities which are reasonably dense urban nodes providing an integrated, mixed-use, walkable environment.

In Paris, officials are looking to implement such measures in the form of a 15-minute city. Under that model, the majority of residents' needs, from work to shopping to leisure activities would all be found within a 15-minute walk or bike ride from their home. These urban villages could ultimately prove successful in the French capital because it has the density needed to make them work. Other cities such as Portland, Detroit, Dallas, and Chicago are all exploring versions of the 15-minute, or in some instances, 20-minute city.

Likewise, Barcelona is already implementing superblocks or collections of nine blocks arranged in a neat three-by-three fashion. Car traffic can freely circulate the perimeter of each superblock, but most vehicles are barred from the interior. As a result, Barcelona is taking space that was devoted to a single-use, car travel and transforming it into space open to multiple uses, including walking, biking, playing, lounging, eating and drinking, etc. In that sense, the superblocks become as much about health, community building and sustainability as they are about alternative modes of mobility.

10.8.2 Outlook

The COVID-19 pandemic has been an experience that we won't ever forget. It has changed our lives in many temporary, and we are sure some permanent, ways. It has also given us as designers a chance to reflect on the way work is done and its relationship to our urban environments. Our pandemic-driven increasing involvement with digital technology helped us to survive, and it won't end when we can eventually stop wearing face coverings to protect our health. We have learned that human interaction can't be replaced by technology, and we have been reminded in undeniable ways that without human health, we can't have economic health. Our world continues to face environmental, social and political challenges, all of which impact our cities. As the pandemic recedes, it is a hopeful time as I see a greater willingness to tackle the difficult problems and find intelligent design solutions to improve the quality of life in our cities.

I am deeply indebted to the Gensler Research Institute and my colleagues Sofia Song, Mark Erdley, Cameron Kraus, Joseph Joseph, Vignesh Kaushik, Andre Brumfield and Carlos Cubillos for their significant contributions to this chapter.

CHAPTER 11

Nature, the Pandemic, and the Resilience of Cities

Case Study of Ottawa, Canada

Adrienne Yuen

Standards Council of Canada, Ottawa, Canada

CONTENTS

DOI: 10.1201/9781003148715-11

11.1 INTRODUCTION

At time of publication, the world has now lived through the COVID-19 pandemic for more than a year. Although a handful of countries are containing the virus, cases are surging in many others. COVID slashed an estimated 20% of world GDP in April 2020 with recovery expected to be long and slow (The Economist, 2021). Vaccines have fortunately been developed and approved far earlier than most expected, with the largest vaccination campaign in history now underway. Yet much uncertainty exists around how the pandemic story will unfold: How long will the virus continue to pose a threat to humans? How can we boost our collective resilience and recover from the economic impacts while the pandemic maintains its grip on the world? How will struggling societies cope with the parallel crisis of climate change?

Much is still unknown but cities, institutions, researchers, civil society and thought leaders around the world are now articulating pathways forward that depart from the unsustainable models of the past. At the heart of many of these visions is the concept of a green recovery and a greater role for nature. If the world's health and environmental problems are symptoms of a dominant species out of balance with nature then the balance must be restored to ensure a sustainable recovery. With the massive disruptions that the pandemic has wrought on cities and the adaptive experiments cities have pursued in response, there are signs that nature can and should be our ally.

This chapter reviews what COVID has revealed about cities, nature, and our relationship with the natural world. In this look back over the pandemic experience so far, the author focuses on three aspects in particular: (1) Increases in park visits in cities across the world, along with evidence of associated mental health, physical health and social benefits; (2) the exacerbation of COVID challenges by climate-change-related disasters; and (3) the spotlight that COVID has thrown on habitat loss, biodiversity loss and other drivers of zoonotic diseases. These three considerations build the case for a renewed focus on nature in how we plan, design and run our cities. The chapter takes the City of Ottawa, Canada as a case study, reviewing how authorities and people have responded to the crisis and identifying lessons learned. It discusses five key considerations in bringing nature back to cities, actions which will also ensure local resilience in the face of current and future crises.

11.2 WHAT COVID-19 HAS REVEALED ABOUT CITIES, NATURE, AND OUR RELATIONSHIP WITH THE NATURAL WORLD

11.2.1 Increases in Park Visits in Cities across the World and Research on the Associated Mental Health, Physical Health and Social Benefits

The coronavirus was first detected in humans in the city of Wuhan, China in December 2019; by the end of January 2020 the World Health Organization (WHO) had declared a global public health emergency. One by one, national governments imposed sweeping lockdowns on whole cities, regions, and countries. Shops and restaurants were shuttered. Schooling was postponed. International borders were shut down. Office workers were told to work from home. Households were urged to shelter in place to stop the spread of the highly contagious, selectively lethal and still little-understood virus.

The effects of these lockdowns were visible and palpable. With vehicles off the streets, residents of many cities breathed noticeably cleaner air and enjoyed bluer skies. Urban dwellers shared photos of wildlife roaming through empty streets. As the massive shift to telework stretched on for weeks and months, business districts sat empty while residential areas became the new loci of activity. When evidence emerged that the coronavirus was less transmissible in outdoor settings, cities loosened rules to permit outdoor activities with physical distancing. People began flocking to parks and other outdoor spaces for recreation, socialisation, and respite. An analysis of data from Google COVID-19 Community Mobility Reports for ten countries across Europe, Asia and the Americas found increases in park visits in all countries between mid-February and late May 2020, against baseline data from the five-week period of 3 January to 6 February 2020 (Geng, Innes, Wu, & Wang, 2020). In Canada, the Google reports showed a steady increase to 80% above baseline by early June, and to 151% by mid-September (Flanagan, 2020). In January 2021, one of the country's coldest months of the year, the national figure was still 19% above baseline (Google, 2021).

Evidence suggested that for many people, nature and the outdoors were a much-needed antidote to the stress of the pandemic. A survey of over 2000 people in 68 countries reported that exercise (81%) and relaxation (72%) were the main reasons provided for seeking outdoor space with errands (52%) as the third reason (O'Connor, 2020). Further, researchers who surveyed people during the pandemic found many health and social benefits of visiting outdoor spaces. In a survey of 386 adults in the city of Chengdu, China in April 2020, nearly all respondents reported a perceived positive impact to their physical health after visiting a park and that the visit allowed them to meet their social interaction needs; indeed the lower the self-reported levels of physical health and social interaction before visiting the park, the greater the post-visit effects (Xie, Luo, Furuya, & Sun, 2020).

These findings corroborate a growing number of studies showing the linkages between exposure to nature and health. A study of 3000 people in Tokyo, Japan found associations between the frequency of green space use with increased levels of self-esteem, life satisfaction and subjective happiness, and decreased levels of depression, anxiety and loneliness; even being able to see greenery outside the window from home improved mental health (Soga, Evans, Tsuchiya, & Fukano, 2020). Research at the University of Manchester found that older people's health is connected to the quantity, quality, and proximity of natural areas (Lindley, et al., 2020). An 11-year study of 1.3 million people living in 30 cities across Canada found that living in greener areas was associated with a 8%–12% reduction in mortality from several common causes of death, including cardiovascular and respiratory diseases (Crouse, et al., 2017).

These benefits however do not accrue equally, due to the uneven access to green space across socioeconomic lines; in other words, wealthier neighbourhoods tend to have access to more green space (Xie, Luo, Furuya, & Sun, 2020; Crouse, et al., 2017; Nesbitt, Meitner, Girling, Sheppard, & Lu, 2019; Gerrish & Watkins, 2018). Much as the pandemic threw into sharp relief the presence of inequality and injustice in other areas of economy and society, the unequal distribution of green space—and their benefits—is another gap regarding nature and cities that the pandemic has brought to the forefront.

11.2.2 COVID and Climate Change: The Exacerbation of COVID Challenges

While the public health crisis upended lives and societies across the globe, climate change has continued its course. Globally, 2020 tied with 2016 for the warmest year on record, while 2011–2020 broke records as the warmest decade (Copernicus Climate Change Service, 2021). Devastating natural disasters and other climate change-related impacts have added to and amplified the pandemic's many challenges worldwide. A record-setting 8.9 million acres burned in the wildfires of the western United States (Center for Disaster Philanthrophy, 2020). Bushfires in Australia caused approximately 500 deaths and killed an estimated 3 billion animals (BBC News, 2020). A record number of hurricanes was recorded by the U.S. National Hurricane Center; so many that it ran through its list of alphabetical names and began using Greek letters (Whang, 2020).

Even as the pandemic rages on, international public concerns around climate change are high. A median of 70% of respondents surveyed across 14 countries in the summer of 2020 said that climate change is a major threat to their country, nearly the same median (69%) as those who said that infectious diseases are a major threat (Fagan & Huang, 2020). The negative impacts of climate change on health, such as the effects of extreme heat on older people and the stress of displacement from natural disaster, are well-documented (The Lancet, 2020). In seeking to recover from COVID, it is clear that societies cannot pursue responses that do not also address the impacts and drivers of climate change.

Nature-based solutions are one area of climate change response that is receiving growing support. In 2020, the Global Commission on Adaptation, an initiative launched by 17 convening countries in The Hague in 2018 published a *Call to Action for a Climate-Resilient Recovery from COVID-19*. The statement calls for accelerated progress in seven areas, including nature-based solutions, citing higher economic returns, greater job creation, faster implementation, and greater long-run sustainability as compared to traditional infrastructure (Global Commission on Adaptation, 2020).

The C40 Cities Mayors' Agenda for a Green and Just Recovery also promotes nature-based solutions as part of an equitable and sustainable recovery plan (C40, 2020). Released by the C40 Global Mayors COVID-19 Recovery Task Force, a group of 12 city leaders from Europe, Africa, Asia, and the Americas, the agenda lays out three key areas of action: Action to create jobs and an inclusive economy, action for resilience and equity, and action for health and well-being. The report argues that nature-based solutions such as parks, green roofs and permeable pavements can be economical and effective responses to extreme heat and natural disaster, while reducing the risks of vector- and water-borne diseases and improving livability and increasing physical and mental health (C40, 2020).

11.2.3 Biodiversity, Habitat Loss and Other Drivers of Zoonotic Diseases

A third reason for putting nature at the heart of economic recovery from COVID is the urgency of the biodiversity crisis. Evidence suggests that COVID originated from a "wet market" in China, one of many in Asia where live animals, including wildlife, are sold (Mizumoto, Kagaya, & Chowell, 2020). This has thrown a spotlight on wildlife trafficking and trade, and our relationship with non-human animals more broadly. According to the WHO, COVID-19 is only the latest of a series of infectious diseases, including HIV/AIDS, SARS and Ebola, to cross the species barrier from wildlife to humans. An estimated 60% of infectious diseases in humans originate this way, stemming from human pressures on wildlife and their habitats such as deforestation, intensive and polluting agricultural practices, and unsafe management and consumption of wildlife (World Health Organization, 2020). Accordingly, the *WHO Manifesto for a healthy recovery from COVID-19* places "Protect and preserve the source of human health: Nature" at the top of its six "prescriptions for a healthy, green recovery" (World Health Organization, 2020).

Governments clearly need to strengthen enforcement to prevent the unsafe handling of wildlife where these zoonotic diseases are likely to emerge but urgent action is needed on many other fronts. In fact, far from giving nature a "break" through a slowdown in global economic activity, COVID has driven an increase in land-grabbing, deforestation, illegal mining, and wildlife poaching in many rural areas in the tropics (Troëng, Barbier, & Rodríguez, 2020). Habitat degradation due to human activity continues to drive biodiversity loss, drawing down the diversity of species that protect ecosystems from the decimating effects of disease—a beneficial phenomenon called the dilution effect (World Health Organization, 2020). Monocultures, employed by the industrial agricultural sector to maximise production, are also highly vulnerable to devastation by disease.

Although protecting nature and carving out more space for it in cities cannot solve the above problems alone, they are a start. The Global Commission on Adaptation notes that intact ecosystems can limit the spread of zoonotic diseases and that investments in nature-based solutions can decrease biodiversity loss, reduce carbon emissions, strengthen resilience to disasters, improve food security, and benefit human and ecosystem health (Global Commission on Adaptation, 2020).

11.2.4 The Case for a Renewed Focus on Nature in Cities

The difficulties of the COVID pandemic have shone a spotlight on the importance of green space, the continued urgency of the climate crisis, and the need to address the biodiversity crisis and fix our relationship with nature. These observations and research on cities, nature, and our relationship with the environment—triggered by an unprecedented year—form a compelling case for changing the way we plan and manage our cities. This chapter now turns to the City of Ottawa as an example of a Canadian response to these developments.

11.3 COVID, NATURE AND CITIES: CASE STUDY OF OTTAWA, CANADA

11.3.1 Increasing Access to Parks and Green Spaces

Canada's capital, the City of Ottawa, is situated in the southern part of the Province of Ontario, on the border with the Province of Québec. With one million people living in an amalgamation of urban, suburban, and rural neighbourhoods, it is the country's sixth largest by population and fourth in terms of land area. As in many other cities, the pandemic brought to Ottawa a mix of disruption, experimentation, and adaptation. It also brought a great deal of stress: Seven months after the initial lockdowns, nearly two-thirds of Ottawa residents reported that COVID had a negative impact on their mental health, citing quarantine, feelings of isolation, and mental health struggles such as anxiety and depression (Nanos, 2020).

Indoor activities were drastically limited by the restrictions of the pandemic but as soon as parks and other outdoor public spaces began to open, residents flocked outside to walk, cycle, hike, and relax. To meet this surge in demand, two authorities with jurisdiction over Ottawa's public spaces – the National Capital Commission and the City of Ottawa – began experimenting with new uses of those spaces to give pedestrians, cyclists, and other active users additional opportunities to safely enjoy the outdoors.

The National Capital Commission (NCC) is a federal crown corporation that owns and manages 536 km² of land in the National Capital Region (NCR), a federally designated area that straddles the provincial border between Ontario and Québec. The NCR includes the City of Ottawa; the neighbouring City of Gatineau, Québec; and surrounding urban and rural communities (Figure 11.1). The NCC was created by Canada's Parliament in 1959 to serve as the Capital Region's main long-term urban planner and to manage

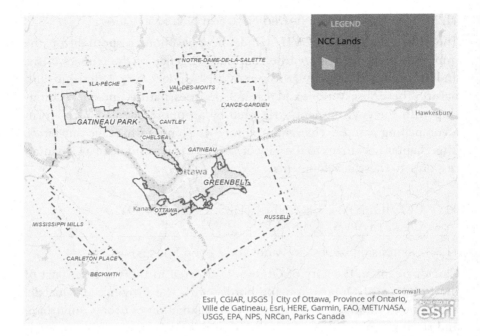

FIGURE 11.1 Map of the National Capital Region. NCC lands are shown in darker grey. The City of Ottawa includes the area south of the Ottawa River (labeled in the image as "Riviere des Outaouais") within the bold dotted line, and includes the Greenbelt. (Credit: National Capital Commission)

the NCR's important cultural, heritage, and natural assets. Natural assets in this diverse portfolio include Gatineau Park (Figure 11.2), a conservation park in Québec covering 361 km²; the Capital Greenbelt, a conservation area surrounding Ottawa's urban core that includes 20,000 hectares of farmland, forests, sand dunes, and wetlands; and a variety of urban parks and green spaces. It is also responsible for 106 km of parkways, over 200 km of multi-use pathways, two interprovincial bridges, the Rideau Canal, and 555 properties for residential, agricultural, institutional, recreational, and commercial purposes (National Capital Commission, n.d.).

The NCC was the first authority in Canada to close some of its parkways to cars regularly for cyclists and pedestrians to enjoy, starting in 1970. During the pandemic, it increased the extent and frequency of these closures to provide additional opportunities for residents of the NCR to enjoy the outdoors. From March to October 2020, the NCC's Parkways for People Initiatives closed over 50% of its parkways to vehicles for various

FIGURE 11.2   View from a trail in Gatineau Park. (Credit: Author)

parts of the week, representing the highest percentage of any jurisdiction in Canada. This included the two major urban east-west roadways during the day on weekends; a 2.5-km stretch along the Rideau Canal every day during the summer; and scenic parkways at the southern end of Gatineau Park nearly all week from spring to fall (Nussbaum, 2020).

The uptake of these additional spaces exceeded expectations. Although roads were "closed," use of the roads was greater than when the parkways were for vehicles only. Over 700,000 visits by active users – cyclists, pedestrians, joggers, in-line skaters, and others – were counted throughout the initiative (Nussbaum, 2020). In an online survey of approximately 8000 visitors, 95% supported the initiative, citing benefits such as a safe space to enjoy the outdoors, to explore new neighbourhoods, and to be active with the family without the usual noise and pollution of vehicles (Nussbaum, 2020).

The NCC has continued to find ways to maintain and expand access to natural and recreational assets, including launching a pilot project to expand its network of groomed winter trails to 100 km, nearly 50% more than the previous year (Figure 11.3). This will allow users to pursue a range of winter activities such as cross-country skiing, snowshoeing, and fat biking (National Capital Commission, 2020). Even as the Province of Ontario entered a second lockdown in December 2020 due to rising COVID cases, the NCC confirmed that the Rideau Canal Skateway would remain open

FIGURE 11.3   A winter trail in the Capital Greenbelt. (Credit: Author)

for physically distanced recreation, to support residents' mental and physical health (CBC News, 2021).

The City of Ottawa itself has also adopted a variety of flexible approaches to allow residents to safely enjoy the outdoors. In the spring of 2020, when it became apparent that outdoor settings posed a lower risk than indoor settings for viral transmission, several city councillors pushed for the closure of certain streets to vehicles so that pedestrians would have more space to walk. As a result, from spring through fall 2020 several main streets were closed to vehicles during certain days and hours of the week (Woods, 2020; Glowacki, 2020; Ottawa Citizen, 2020). In other places, sidewalks were widened to include part of the roadway, which allowed pedestrians to use them while following physical distancing and still permitting traffic to flow through (Lord, 2020). As the pandemic stretched into the winter of 2020–2021, the City also implemented capacity limits at community outdoor skating rinks and sledding hills, required online reservations for its refrigerated rinks, and deployed by-law officers to monitor these areas (City of Ottawa, 2021).

11.3.2   Green Spaces, Climate Change and Biodiversity

Green and outdoor spaces have been firmly in demand during the pandemic, as the NCC's experience and community mobility data for Canada have shown. They have also worked double duty on the ecological front.

According to a study commissioned by the NCC, communities in the NCR receive benefits worth an average of CAD $332 million per year from the NCC's green spaces, or CAD $5 billion over 20 years. These include direct monetary benefits such as wood and agricultural products, as well as a range of ecological benefits including air quality control, water filtration, climate regulation, carbon storage, wildlife habitat, and erosion control (NCC).

Biodiversity corridors in the NCR play an important role in supporting local flora and fauna. An NCC multi-use pathway (Figure 11.4) runs through the Lac Deschênes forests, enabling access to nature for cyclists, pedestrians, and other active users. The forests are in turn part of the Lac Deschênes-Ottawa River Important Bird Area (IBA), a 45-km-long zone encircling the Ottawa River that runs between Québec and Ontario, and the river's adjacent lands. An IBA is designated as such if, using standardised criteria and empirical evidence, it is shown to support certain thresholds of globally threatened species of birds, large groups of birds, and/or birds restricted by range or habitat (IBA Canada, n.d.). The Lac Deschênes-Ottawa River IBA serves as an important stopover point for migratory birds on intercontinental journeys, such as brant, Canada geese, swallows, and chimney swifts; and is home to several species of gulls, the Black-crowned Night-heron, and the Great Egret (IBA Canada, n.d.).

FIGURE 11.4   A National Capital Commission multi-use pathway. (Credit: Author)

11.3.3 Calls for an Integrating Nature in Canada's COVID Recovery

The experience of the NCR during COVID-19, and continued moves to expand access to green and outdoor spaces, is in line with a growing call among Canadian municipalities and thought leaders for policies and funding arrangements from provincial and federal governments that support cities in leading the green recovery, including creating a greater role for nature in cities. The *2020 Declaration for Resilience in Canadian Cities*, for example, led by former City of Toronto Chief Planner Jennifer Keesmaat, calls for the responsible use of land, including stronger restrictions on urban sprawl as well as a detailed funded plan to achieve a 40% urban tree canopy (Keesmaat, 2020).

The Federation of Canadian Municipalities (FCM) an organisation that represents nearly 2000 municipalities of all sizes also recognises the importance of natural infrastructure, parks, and protected areas to improve residents' access to nature, support biodiversity, and improve overall community resilience against climate change. Among other recommendations, the FCM recommends expanding federal funding to cities to enable municipalities to acquire land for parks, protected areas, and natural infrastructure; and eliminating the current CAD $20-million project eligibility floor to allow natural infrastructure projects and smaller communities to access funding (Federation of Canadian Municipalities, 2020).

The Task Force for Climate Resilience, an independent group of Canadian finance, policy, and sustainability leaders, asserts that "our nature economy is Canada's secret weapon for spurring a resilient recovery from COVID-19" (Task Force for Resilient Recovery, 2020). The group calls for investments in natural infrastructure, an expansion of Canada's Protected Areas Network, policies to stimulate private investment in nature-based services, and funding to create employment opportunities in the nature economy (Task Force for Resilient Recovery, 2020).

11.4 CONSIDERATIONS

The above Canadian voices join a global chorus of researchers and thought leaders calling for the embedding of nature in COVID recovery measures. Increases in visits to green space in cities, evidence on the mental and physical health benefits of green exposure, and continued concerns over climate change and biodiversity loss suggest broad public support for such an agenda. Advancing this agenda will need to include several considerations.

11.4.1 Quality and Variety

Not all green spaces are equal. They can range from highly built, serviced, cultivated, manicured, and maintained spaces; to wild places with little human-built infrastructure. They also vary greatly in size and in proximity to developed urban areas. A key benefit of these wilder spaces is the opportunity for "soft fascination," or a state of relaxed curiosity (Kaplan & Kaplan, 1989). Fitzgibbons calls such places Restorative Natural Areas which generally have the following characteristics and features: Quiet, with minimal crowding; limited to no road noise; minimal built infrastructure; natural elements like water, large trees, wildflowers; unlikely to be a pocket park, green roof, sport field, or playground; and a source of co-benefits for conservation and habitat restoration (Fitzgibbons, Not All Green Space is Created Equal—or Equally Accessible to All, 2020). She writes:

> A restorative environment usually has four main elements: soft fascination; a sense of being away or 'escaped' from the usual setting; a setting that is satisfying to the individual's intent or purpose in being there (people feel that they can get what they went there for); and, a sense of perspective or being a part of a "larger whole." (Fitzgibbons, Access to Nature in Vancouver: What does it mean, and can we map it?, 2020)

These criteria help design or redesign green spaces and are important outcomes to aim for alongside other benefits such as flood mitigation or carbon sequestration. This is not to imply, however, that more managed or structured green spaces are less valuable; they simply serve other purposes. An online survey of park users in the Province of British Columbia and elsewhere, Fitzgibbons finds that most people would like to see a diversity of green spaces in a city, with relatively more managed and accessible parks as well those that are left more wild (Fitzgibbons, Access to Nature in Vancouver: What does it mean, and can we map it?, 2020).

11.4.2 Access and Equity

The pandemic has also exposed the fault lines of socioeconomic inequality, including unequal access to green space depending on the neighbourhood one lives in, or the type of dwelling they inhabit. During the early months of the pandemic, researchers found that mental and physical health were the main drivers of public space use; it also found that people tended to use spaces closer to home, which points to the need to ensure green spaces are equitably distributed (O'Connor, 2020). Plenty of evidence also exists for the links between nature and mental health. For this reason, green infrastructure should also be considered a form of psychological infrastructure (Tobin Garrett & Stark, 2020).

It is also important to recognise that greening a city does not automatically lead to equitable outcomes. Pursued in the absence of equity or justice considerations, adding green space can further socioeconomic inequality, for instance by driving up nearby property values and pushing lower-income people out of neighbourhoods. New York City is one such example, where urban revitalisation in Prospect Park, along the High Line, and in several Brooklyn neighbourhoods raised housing prices and forced out low- and middle-income residents (Grinspan, Pool, Trivedi, Anderson, & Bouyé, 2020). Efforts to introduce green space must be pursued in meaningful collaboration with the community, informed by the needs of diverse users, and consider any potentially negative impacts on underserved or underrepresented demographics to avoid creating or exacerbating socioeconomic inequalities.

11.4.3 Valuation

An ongoing challenge of protecting nature is valuing access to rather than consumption of green space; it is easy to calculate how much an area of land is worth developed but not as easy to calculate its worth undeveloped. The benefits of green infrastructure are still not widely captured in accounting or asset management practices. Organisations such as the Municipal Natural Assets Initiative (MNAI) are filling this gap by working with Canadian municipalities to identify, value, and account for natural assets in financial planning and asset management. MNAI began its work with the Town of Gibsons, British Columbia, a small community with limited resources for infrastructure maintenance and replacement. MNAI assisted Gibsons in using natural capital management and green infrastructure to improve its resilience to climate change impacts (Municipal Natural Assets Initiative, 2020). The Credit Valley Conservation Authority (CVCA) in Ontario is another organisation that is leading efforts to put a robust price on climate resilience interventions. CVCA has worked with software developers to develop a Risk and Return on Investment Tool that allows users to understand by inputting and manipulating multiple layers of mapping data, the areas of greatest flood, and erosion risk in a community; as well as the quantified costs and benefits of different solutions such as green infrastructure (Risk Sciences International, 2021).

Even forward-thinking organisations such as the NCC are not immune to the pressure to develop greenfield. Although it has not yet been finalised, the NCC has proposed to rezone 3.7 hectares of urban greenspace to build six foreign embassies as well as the same number of accompanying parking lots. If the proposal is approved, 66 of approximately 130 mature trees would be felled, removing habitat and pathways for 63 varieties of birds. In addition to this loss of wildlife habitat, it would also represent a significant decline in green space per resident, particularly as the ward in which it is located is forecast to see up to 400% intensification in coming years through the development of residential properties (Hintonburg Community Association Inc., 2021).

11.4.4 Population Growth and Development Pressures

By 2050, 68% of people will live in cities (UN, 2018). In Canada, where approximately 80% of the population are urban dwellers, the country's population will rise from 34 million today to nearly 49 million in 2050 (Globe and Mail, 2020). The cities that will absorb them will need to provide housing, transit, infrastructure, schools, grocery stores, and other amenities. Just as existing spaces will need to be redeveloped, green spaces will come under even greater pressure for new development. Indeed following a May 2020 decision by Ottawa City Council to expand the urban boundary, Ottawa city planners are now considering the development of 1,011 hectares of greenspace in the municipality's rural areas (Porter, 2021).

Canadians overall value nature, but due to permissive urban policies and a culture of car-dependency, new development is still overwhelmingly land-inefficient; approximately three-quarters of new housing in the last decade was built in sprawling suburban communities (Keesmaat, 2020). This tendency is harmful not only for the conservation of natural areas but for the development of sustainable, livable communities. With the idea of the "15-minute city" gaining traction with urban planners around the world (Reguly, 2020), the continuation of low-density development is a backward trend.

To avoid locking in further car-dependent communities with large environmental footprints, cities need to adopt long-term planning approaches that prioritise appropriate levels of urban density, acknowledge and quantify the value of existing natural assets and establish protections for those assets. Planners and decision-makers can learn about successful approaches from other jurisdictions. For example, Scotland's Scottish Planning Policy (2010) recommends the creation of "Outdoor Access Plans" for any proposed development that might impact demand or access to a given green space or inland water. These plans establish a baseline of current accessibility and demand for the natural area, and project the impact that the project might have. This increases the transparency of the impacts of a development and involves the developer in creating solutions to maintaining or improving access to nature for existing and new residents (Fitzgibbons, Access to Nature in Vancouver: What does it mean, and can we map it?, 2020).

11.4.5 Public Engagement and Support

Education and fostering an appreciation for nature is also an important component in integrating nature in cities. Just as a nature interpreter broadens a visitor's understanding and appreciation of a natural destination, education and outreach encourage city residents to experience the many benefits of nature and invite them to be stewards of their community's natural assets. The Biophilic Cities Network which comprises 24 partner cities across Europe, Asia, the Americas, and Australasia, describes the biophilic city in terms of the physical presence of nature in the community but also the connection that residents have with it:

> While the question of what constitutes a biophilic city is an open and evolving one, it is a city that contains abundant biodiversity and nature, that works to conserve that nature as well as creatively insert new forms of nature, and fosters connections to the natural world … A biophilic city places nature at the core of its design and planning, and works to create abundant opportunities for people to learn about and connect with this nature. A biophilic city understands and celebrates that its role is to provide habitat for many different forms of life, and advocate for humane co-existence. Biophilic cities protect, grow, and celebrate local nature, but also work on behalf of nature beyond their borders.
>
> *(Biophilic Cities, 2015)*

To be a partner in the network, a city is required to select or establish at least one indicator from five categories, one of which is "biophilic engagement, participation, activities and knowledge." Examples under this category include the ability of city residents to identify common species of flora or fauna, extent of membership in nature or outdoor clubs. Another category relates to institutions, planning, and governance, including the number of pupils exposed to nature education or the number of city schools with eco- or bio-literacy curricula (Biophilic Cities, 2015).

The Singapore Biodiversity Index, a global standard developed to help cities establish a baseline on biodiversity and then measure its progress over time, also includes indicators relating to education and awareness. Cities can score points for the level of coverage of biodiversity in the local school curriculum and for the number of outreach or public awareness events held in a year (User's Manual on the Singapore Index on Cities' Biodiversity (also known as the City Biodiversity Index), 2014).

11.5 CONCLUSION

The world is grappling with multiple crises: The COVID-19 pandemic, the climate crisis and the biodiversity crisis. COVID-19 has upended daily life across the globe. It has spurred an increase in visits to parks and other outdoor spaces, highlighting for many people the value of green spaces to mental and physical health. Meanwhile, climate change is causing immense destruction, and human activities are steadily drawing down our collective biodiversity bank. These realities have prompted researchers, prominent institutions, and thought leaders to call for a nature-based recovery from COVID.

The restoration of nature within urban areas will not, on its own, solve these problems. But with over two-third of people expected to be urban dwellers by mid-century, lessening the environmental footprint of cities must be part of the solution. Bringing nature back to cities (Figure 11.5) and ensuring that everyone has access to nature will make us happier and healthier and live longer lives. Nature will strengthen our communities against climate change, often for much cheaper than traditional engineered structures can accomplish. It will bring back biodiversity and strengthen the ecosystems upon which we ultimately depend.

The trajectory of the pandemic is still uncertain. Regardless of what unfolds, protecting and enhancing the variety of green spaces within and near urban communities brings a myriad of benefits. Guided by scientific evidence and implemented with an equity lens, it is a "no-regrets" approach that will enhance health and livability, and help communities weather these challenges and others for decades to come.

FIGURE 11.5   Urban view of the Ottawa River from a multi-use pathway. (Credit: Author)

REFERENCES

BBC News. (2020, July 28). *Australia's fires 'killed or harmed three billion animals'*. Retrieved from https://www.bbc.com/news/world-australia-53549936#:~: text=Nearly%20three%20billion%20animals%20were,)%2C%20which%20 commissioned%20the%20report.

Biophilic Cities. (2015, June). *Guidelines for participation in the biophilic cities network: Expectations and submittal requirements for partner cities.* Retrieved from https://www.biophiliccities.org/s/BiophilicCitiesNetworkGuidelines. pdf.

C40. (2020). *C40 Mayors' Agenda for a green and just recovery.* C40 Cities. Retrieved from https://c40-production-images.s3.amazonaws.com/other_ uploads/images/2093_C40_Cities_%282020%29_Mayors_Agenda_for_a_ Green_and_Just_Recovery.original.pdf?1594824518.

CBC News. (2021, January 15). *The Rideau Canal Skateway will open during the stay-at-home order. Here's why.* Retrieved from https://www.cbc.ca/news/ canada/ottawa/rideau-canal-staying-open-heres-why-1.5873582.

Center for Disaster Philanthropy. (2020, December 7). *2020 North American wildfire season.* Retrieved from https://disasterphilanthropy.org/disaster/ 2020-california-wildfires/.

City of Ottawa. (2021, January 7). *City implements 25-person limit at rinks and sledding hills, reservation system for refrigerated rinks.* Retrieved from https://ottawa.ca/en/news/city-implements-25-person-limit-rinks-and-sledding-hills-reservation-system-refrigerated-rinks.

Copernicus Climate Change Service. (2021, January 8). *Copernicus: 2020 warmest year on record for Europe; globally, 2020 ties with 2016 for warmest year recorded.* Retrieved from https://climate.copernicus.eu/copernicus-2020-warmest-year-record-europe-globally-2020-ties-2016-warmest-year-recorded.

Crouse, D., Pinault, L., Balram, A., Hystad, P., Peters, P., Chen, H., Villeneuve, P. (2017, October 1). Urban greenness and mortality in Canada's largest cities: a national cohort study. *The Lancet Planetary Health, 1*(7), E289–E297.

Fagan, M., & Huang, C. (2020, October 16). *Many globally are as concerned about climate change as about the spread of infectious diseases.* Retrieved from Pew Research Center: https://www.pewresearch.org/fact-tank/2020/10/16/many-globally-are-as-concerned-about-climate-change-as-about-the-spread-of-infectious-diseases/.

Federation of Canadian Municipalities. (2020, November). *Building back better: Municipal recommendations for Canada's post-COVID recovery.* Retrieved from https://data.fcm.ca/documents/COVID-19/fcm-building-back-better-together.pdf.

Fitzgibbons, J. (2020, July). *Access to nature in Vancouver: What does it mean, and can we map it?* Retrieved from https://drive.google.com/file/d/11KatuByGn1ZMn0gyost-lL4kxtNCjRaI/view.

Fitzgibbons, J. (2020, November 11). *Not all green space is created equal—or equally accessible to all.* Retrieved from The City Fix: https://thecityfix.com/blog/not-all-green-space-is-created-equal-or-equally-accessible-to-all/.

Flanagan, R. (2020, September 20). *Canadians are still flocking to parks and businesses as country braces for second wave.* Retrieved from CTV News: https://www.ctvnews.ca/health/coronavirus/canadians-are-still-flocking-to-parks-and-businesses-as-country-braces-for-second-wave-1.5112544.

Geng, D., Innes, J., Wu, W., & Wang, G. (2020, November 12). Impacts of COVID-19 pandemic on urban park visitation: a global analysis. *Journal of Forestry Research.* Retrieved from https://link.springer.com/article/10.1007/s11676-020-01249-w.

Gerrish, E., & Watkins, S. (2018). The relationship between urban forests and income: A meta-analysis. *Landscape and Urban Planning,* 293–308. Retrieved from https://urbanforestry.indiana.edu/doc/publications/2018-watkins-UF-income.pdf.

Global Commission on Adaptation. (2020, July 9). *Call to action for a climate-resilient recovery from COVID-19.* Retrieved from https://cdn.gca.org/assets/2020-07/Global_Commission_Adapation_COVID_Resilience_Statement.pdf.

Globe and Mail. (2020, January 6). Retrieved from Canada's cities are about to add millions of new residents. They can't all drive to work.

Glowacki, L. (2020, April 9). *Westboro street closes to free up space for pedestrians, cyclists.* Retrieved from CBC News: https://www.cbc.ca/news/canada/ottawa/byron-avenue-closes-1.5528620.

Google. (2021, January 10). *COVID-19 Community Mobility Report: Canada: January 10, 2021.* Retrieved from https://www.gstatic.com/covid19/mobility/2021-01-10_CA_Mobility_Report_en.pdf.

Grinspan, D., Pool, J., Trivedi, A., Anderson, J., & Bouyé, M. (2020, October 5). *Green Space: An Underestimated Tool to Create More Equal Cities.* Retrieved from The City Fix: https://thecityfix.com/blog/green-space-an-underestimated-tool-to-create-more-equal-cities/.

Hintonburg Community Association Inc. (2021, January 19). *Re: Withdrawal of Applications D01-01-19-001 and D02-02-19-0072.* Retrieved from http://hintonburg.com/wp-content/uploads/2021/02/HCA-letter-Opposing-Embassy-Proposal-.pdf.

IBA Canada. (n.d.). *What is an important bird area?* Retrieved from https://www.ibacanada.org/iba_what.jsp?lang=en.

Kaplan, R., & Kaplan, S. (1989). *The Experience of Nature: A Psychological Perspective.* Cambridge University Press.

Keesmaat, J., et al. (2020). *2020 Declaration for resilience in Canadian cities.* Retrieved from https://www.2020declaration.ca.

Lindley, S., Ashton, J., Barker, A., Benton, J., Cavan, G., Christian, R., ... Wossink, A. (2020). *Nature and ageing well in towns and cities: Why the natural environment matters for healthy ageing.* Retrieved from https://ghiadotorgdotuk.files.wordpress.com/2020/01/ghia_report_online_hires.pdf.

Lord, C. (2020, April 15). *Coronavirus: Some Ottawa roads set to open to pedestrians.* Retrieved from Global News: https://globalnews.ca/news/6820441/coronavirus-ottawa-roads-open-to-pedestrians/.

Mizumoto, K., Kagaya, K., & Chowell, G. (2020). Effect of a wet market on coronavirus disease (COVID-19) transmission dynamics in China, 2019–2020. *International Journal of Infectious Diseases: IJID: Official publication of the International Society for Infectious Diseases, 97,* 96–101. Retrieved from https://www.ncbi.nlm.nih.gov/pmc/articles/PMC7264924/.

Municipal Natural Assets Initiative. (2020). *Where it all started: Town of Gibsons.* Retrieved from https://mnai.ca/key-documents/.

Nanos. (2020, December 6). *A majority of Ottawa residents say the pandemic has had negative effects on their mental health; two thirds show support for a universal basic income program.* Retrieved from https://secureservercdn.net/198.71.233.47/823.910.myftpupload.com/wp-content/uploads/2020/10/2020-1662-Ottawa-Mission-Populated-Report-With-tabs.pdf.

National Capital Commission. (2020). *Groomed multi-use winter trail network.* Retrieved from https://ncc-ccn.gc.ca/places/groomed-multi-use-winter-trails.

National Capital Commission. (n.d.). *Discover Canada's Capital Region.* Retrieved from https://ncc-ccn.maps.arcgis.com/apps/MapJournal/index.html?appid=2cd22a1ad6014313855880f69b5690d7.

National Capital Commission. (n.d.). *Natural Capital: The economic value of NCC green spaces.* Retrieved from https://ncc-ccn.gc.ca/news/natural-capital-the-economic-value-of-ncc-green-spaces-1.

Nesbitt, L., Meitner, M., Girling, C., Sheppard, S., & Lu, Y. (2019). Who has access to urban vegetation? A spatial analysis of distributional green equity in 10 US cities. *Landscape and Urban Planning, 181,* 51–79. Retrieved from https://doi.org/10.1016/j.landurbplan.2018.08.007.

Nussbaum, T. (2020, December 22). *COVID showed us ways we can make our cities more liveable.* Retrieved from National Capital Commission: https://ncc-ccn.gc.ca/news/covid-showed-us-ways-we-can-make-our-cities-more-liveable.

O'Connor, E. (2020, May 7). *Public space plays vital role in pandemic.* Retrieved from Gehl Blog: https://gehlpeople.com/blog/public-space-plays-vital-role-in-pandemic/.

Ottawa Citizen. (2020, July 4). *Somerset Village closes to weekend traffic, Downtown BIA Bank plans weekend vehicle ban.* Retrieved from Ottawa Citizen: https://ottawacitizen.com/news/local-news/somerset-village-closes-to-weekend-traffic-to-encourage-pedestrians-expand-patios.

Porter, K. (2021, January 16). *City staff propose 'gold belt' to hem in future Ottawa development.* Retrieved from CBC News: https://www.cbc.ca/news/canada/ottawa/urban-boundary-map-rural-properties-1.5874871.

Reguly, E. (2020, November 19). *Bikes, pedestrians and the 15-minute city: How the pandemic is propelling urban revolutions.* Retrieved from Globe and Mail: https://www.theglobeandmail.com/canada/article-bikes-pedestrians-and-the-15-minute-city-how-the-pandemic-is/.

Risk Sciences International. (2021). *Risk and Return on Investment Tool (RROIT): Helping Helping municipalities understand their flood risk under current and future climate change scenarios.* Retrieved from https://risksciences.com/rroit/.

Scottish Planning Policy, (2010): Retrieved from https://www.gov.scot/publications/scottish-planning-policy/.

Soga, M., Evans, M., Tsuchiya, K., & Fukano, Y. (2020, November 17). A room with a green view: the importance of nearby nature for mental health during the COVID-19 pandemic. *Ecological Applications.* Retrieved from https://doi.org/10.1002/eap.2248.

Task Force for Resilient Recovery. (2020, September). *Bridge to the Future: Final Report from the Task Force for a Resilient Recovery.* Retrieved from International Institute for Sustainable Development: https://www.iisd.org/sustainable-recovery/news/resilient-recovery-canadian-task-force-shared-final-report/.

The Economist. (2021, January 13). *What is the economic impact of the latest round of lockdowns?* Retrieved from https://www.economist.com/finance-and-economics/2021/01/13/what-is-the-economic-impact-of-the-latest-round-of-lockdowns.

The Lancet. (2020, December 2). Climate and COVID-19: converging crises. *The Lancet,* 397(10269), P71. Retrieved from https://www.thelancet.com/journals/lancet/article/PIIS0140-6736(20)32579-4/fulltext.

Tobin Garrett, J., & Stark, A. (2020). *The Canadian City Parks Report.* Toronto: Park People. Retrieved January 12, 2021, from https://dev.ccpr.parkpeople.ca/uploads/ccpr_print2020_EN_G2_1a87af8874.pdf.

Troëng, S., Barbier, E., & Rodríguez, C. (2020, May 21). *The COVID-19 pandemic is not a break for nature – let's make sure there is one after the crisis.*

Retrieved from World Economic Forum: https://www.weforum.org/platforms/covid-action-platform/articles/covid-19-coronavirus-pandemic-nature-environment-green-stimulus-biodiversity.

UN. (2018, May 16). *68% of the world population projected to live in urban areas by 2050, says UN.* Retrieved from https://www.un.org/development/desa/en/news/population/2018-revision-of-world-urbanization-prospects.html#:~:text=News-,68%25%20of%20the%20world%20population%20projected%20to%20live%20in,areas%20by%202050%2C%20says%20UN&text=Today%2C%2055%25%20of%20the%20.

User's Manual on the Singapore Index on Cities' Biodiversity (also known as the City Biodiversity Index). (2014). Retrieved from Convention on Biological Diversity: https://www.cbd.int/doc/meetings/city/subws-2014-01/other/subws-2014-01-singapore-index-manual-en.pdf.

Whang, O. (2020, September 21). *We've run out of hurricane names. What happens now?* Retrieved from https://www.nationalgeographic.com/science/2020/09/weve-run-out-of-hurricane-names-what-happens-now/.

Woods, M. (2020, July 8). *Bank Street downtown closing to vehicles on Saturdays.* Retrieved from CTV News: https://ottawa.ctvnews.ca/bank-street-downtown-closing-to-vehicles-on-saturdays-1.5015913.

World Health Organization. (2020, May 26). *WHO Manifesto for a healthy recovery from COVID-19.* Retrieved from https://www.who.int/news-room/feature-stories/detail/who-manifesto-for-a-healthy-recovery-from-covid-19.

Xie, J., Luo, S., Furuya, K., & Sun, D. (2020, August 20). Urban parks as green buffers during the COVID-19 pandemic. *Sustainability, 12*(17). Retrieved from https://www.mdpi.com/2071-1050/12/17/6751.

Future of Learning and Teaching in Higher Education Post-COVID-19

Caroline Wong[1], Esther Fink[1] and Abhishek Bhati[1]

[1]James Cook University Singapore, Singapore

CONTENTS

DOI: 10.1201/9781003148715-12

12.1 INTRODUCTION

The year 2020 was a definitive moment as the world of learning and teaching changed tremendously as a result of the COVID-19 pandemic. With social distancing and strict measures to deter face-to-face interactions, many educational institutions pivoted from physical location-based learning to online with the adoption of various digital tools and practices to deliver learning to their learners. The pandemic forced education institutions to innovate in teaching, learning processes and learner management and reimagine their approach to education. Today, learners are facing a similar transition to a less familiar online environment and limited in-person interactions conducted in purpose-built physical environments. The need for digital literacy became paramount as the evolving online educational experiences were manifestations of crisis management necessitating digital citizenship in the post-COVID-19 world (Buchholz et al., 2020). As these realities sink in, stakeholders are evaluating implications on the future of learning and teaching in higher education post-COVID-19.

Technology has profoundly changed the world of higher education and broadened access to information. The MOOC (Massive Open Online Courses) movement took the world by storm chalking up 120 million learners (excluding China) in 2019 with over 2500 courses, 11 online degrees and 170 micro-credentials (Shah, 2019). While there is much debate about what makes for successful online learning experience, there is agreement that technology per se is not sufficient as it is also about how one successfully engages and socialises in the learning journey (Anderson, 2008; Laurillard, 2002; Moore & Kearsley, 2011; Reidsema et al., 2017). By April of 2020, the COVID-19 pandemic transformed education completely to the effect that online learning became a necessity and not a choice.

With this sudden shift away from the face-to-face classroom environment, new challenges surfaced relating to digital competence, confidence and accountability (Passey et al., 2018) in learning, teaching and technological readiness that made problems of equity and access more apparent than ever (Buchholz et al., 2020). Some are wondering whether the adoption of online learning will continue to persist post pandemic and how such a shift might impact the worldwide education market (Li & Lalani, 2020).

12.2 HIGHER EDUCATION IN TRANSITION

Prior to the COVID-19 pandemic, the online learning platform at James Cook University (JCU) Singapore was used across the board but largely as a repository for information, learning materials, assessment and assessment feedback. Learning and Teaching was delivered in face-to-face settings. With the shift to fully online, all previous face-to-face interactions had to shift online giving rise to the questions: How to create a seamless experience for students and faculty, taking into consideration the people and technology side of things? What is the effectiveness of online delivery across time, technology and social setting?

12.3 JCU SINGAPORE STORY

During 2020, while many universities worldwide took to suspending education programmes during the early part of the COVID-19 pandemic, JCU Singapore chose to respond early and decisively, moving all studies online in February 2020 as a pre-emptive measure. For JCU Singapore, the business side of education remained consistently strong throughout the crisis. The majority of the subjects were offered as planned and enrolment numbers did not change drastically and students successfully completed their studies. Our retention rates are in the upper top tenth percentile. The following sections detail the JCU Singapore approach that delivered positive outcomes during challenging times.

12.4 DIGITAL TRANSFORMATION OF EDUCATION

The current pandemic forces explicit adaptation to innovative ways of learning and teaching. At JCU Singapore, it brought together staff from across the various departments to find solutions to the evolving educational landscape. Within weeks, we had shifted from face-to-face to fully online learning and teaching. Students and faculty were facing similar transitions to a less familiar online environment and limited in-person interactions in a purpose-built physical environment.

Faculty being the facilitators of learning, carry greater responsibility and duty of care to ensure opportunities to students to succeed. There is an added impetus for the student to be able to navigate the digital environment while struggling with transitioning to the university setting. The need for digital literacy becomes pertinent in the digital online environment and digital citizenship encompasses the ability to read, write and interact on/across screens to engage with diverse online communities and to work collectively for equity and change apart from having technical skills (Buchholz et al., 2020).

Having a sense of belonging and identity to the institution is essential for both faculty and students. The culture of the institution and the pattern of social interaction within exert a formative effect on the "what" and "how" of learning (Daniels, 2012, p. 2). Some areas of consideration include the interactions between students and faculty, the learning and teaching experience and the mental and emotional well-being of all concerned. The Community of Inquiry (COI) framework developed by Garrison et al. (1999) and the Community of Practice (COP) model by Wenger (1998) hold relevance in shedding light on some of these interactions and experiences.

It is within this context that this chapter addresses the following research questions:

1. What are the realities (benefits and challenges) of digital transformation of education for higher education post-COVID-19?

2. What are the skills and capabilities needed for educators and learners to successfully adapt to the changing environment?

3. What are the implications for the future of learning and teaching in higher education post-COVID-19?

12.5 METHODOLOGY

Using auto-ethnography, this study adopts a reflective approach to examining the realities of higher education post-COVID-19 by using the researchers' own first-hand personal experiences of navigating the pandemic as a rich source of data (Roy & Uekusa, 2020). In this unique and challenging time, we reflect on the journey undertaken by JCU Singapore to cope and deal with the unprecedented crisis. We also examine the impact of the pandemic on learning and teaching and the implications for the kind of knowledge, capabilities and skills required by students to be prepared for the future of work. This research also uses a qualitative methodology through an integrative review of literature drawn from secondary data many of which are sources from discipline-specific, peer-reviewed journals such as *Academy of Management Learning & Education, American Educational Research Journal, British Journal of Educational Technology, Higher Education, Internet and Higher Education, Education Research Review* and *Studies in Higher Education* amongst others.

12.6 REALITY OF DIGITAL TRANSFORMATION OF EDUCATION FOR HIGHER EDUCATION POST-COVID-19

The use of educational technology prior to the COVID-19 pandemic was largely a matter of choice for universities to differentiate themselves from other education providers. While there is a long-standing history of distance learning and online learning, institution-wide adoption are few and far. Many places still largely focused on information sharing such as lecture notes and sometimes lecture recordings (Nordmann et al., 2019; O'Callaghan et al., 2017).

At institutions across the globe, interactive learning experiences in online spaces or online assessment and feedback were limited to a targeted audience, innovators and early adopters. In the new normal, the world is not like it was before. During the COVID-19 pandemic, majority of faculty needed to quickly get familiar with their respective online teaching environment due to social distancing measures and working from home requirements coming into effect at many institutions (Brown & Salmi, 2020).

The new learning and teaching environment is one where synchronous and asynchronous activities and practices exist side by side. In synchronous learning which happens in real time where students and instructor interact in a specific virtual place at a set time, time zone differences can create conflicts in student schedules. This means that there are more complexities and challenges involved with staging these two types of delivery. These conflicts may lead to frustration and cause learners to feel dissatisfied and disconnected from the learning environment (Chundur & Prakash, 2009; Falloon, 2011; Hrastinski, 2008).

Social distancing and travel/movement restrictions have impact on learning and teaching. This is where the need for social presence is pertinent where participants feel affectively connected one to another through ways to connect, communicate, and share their opinions (Swan et al., 2009). Together with the cognitive and teaching presence as suggested by the COI framework (Garrison et al., 1999), these patterns of social interaction enable deep learning to take place. This model assists in the organisation of online and blended educational experiences where students are accountable for their learning and how they learn it (Garrison et al., 1999) resulting in "deep and meaningful learning" (Rourke & Kanuka, 2009, p. 23) as they develop a sense of identity and belonging to the community they are engaged with. Likewise, the COP model by Wenger (1998) suggests that learning occurs through a situated process of participation and socialisation. The next section describes COI adopted in JCU Singapore.

12.7 COMMUNITY OF INQUIRY

Online learning experience does not exist in a vacuum. It is shaped by a community of instructors, learners, technical support, digital tools and practices required to interact and engage with the learning. Such kind of participation and engagement by various stakeholders within an organisation occur in socially situated practices "supported by skills, strategies, and stances that enable the representation and understanding of ideas using a range of modalities enabled by digital tools" (O'Brien & Scharber, 2008, pp. 66–67). As such, learning is shaped by the community activities supported by a shared culture and one in which the learner makes sense of learning based on the experiences and social interactions (Seixas, 1993; von Glasersfeld, 1989). The creation of a social space through these social relations enables individuals to make meaning of each other and the social world (Lefebvre et al., 1991). Garrison et al. (1999) classify these experiences and interactions using the COI framework, which consists of three primary components: cognitive presence, teaching presence and social presence as shown in Figure 12.1.

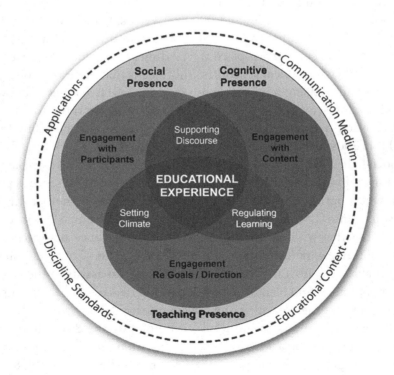

FIGURE 12.1   COI framework (Garrison et al., 1999). (Graphic from: http://www.thecommunityofinquiry.org/coi)

The **COI framework** is consistent with the constructivist approaches to learning in higher education where students would assume responsibility to actively construct and confirm meaning (Garrison, 2007; Swan et al., 2009)

Cognitive presence is the ability of students to reflect, construct meaning and develop understanding through engaging with the assignments, discussions and course tasks for deep learning to take place (Watts, 2017). When students successfully exhibit cognitive presence in an online course, they are actively engaged in creating meaning and confirming their understanding of complex concepts (Garrison & Cleveland-Innes, 2005). At JCU Singapore, asynchronous activities such as readings, pre-recorded lectures and video content, lecture notes, online discussion boards or social media platforms and assessments enable learners to access these materials on their own schedule, so long as they meet the expected deadlines. This encourages self-directed learning where learners engage, reflect and delve deeper into the materials. In synchronous teaching, the use of digital tools like interactive html-5 packages, pen tablets, interactive whiteboards, discussion boards and polling platforms are utilised to promote active learning and reflection.

Social presence is the degree to which participants feel affectively connected one to another through ways to connect, communicate and share their opinions (Swan et al., 2009). At JCU Singapore, the sense of community was enhanced through the interactive breakout and discussion groups where learners learn from their peers. For instance, group work regularly provides opportunities for team members to learn from each other regularly thereby fostering social presence as peer continuity increases comfort and aids group cohesion (Akyol et al., 2009). Ice-breaker activities and opportunities to share personal profiles and session check-ins enable both faculty and peer members of the COI to familiarise themselves with other community members thereby creating a sense of identity and community.

Teaching presence refers to the structuring of the lesson to support both the social and cognitive presence to meet the intended learning outcomes (Tan et al., 2020). One of the key components of teaching presence is the setting and managing of both individual and collaborative learning activities that elicit students' personal experiences and provide opportunities for them to negotiate meaning and the diverse understanding of other members of the community (Anderson, 2018). At JCU Singapore, some of these activities include getting student teams to discuss the mini case studies in online discussion groups or breakout rooms and the tutor acting as a facilitator ensuring that students are actively contributing to the discussion forums. Students also took part in fun quizzes via the interactive

polling platforms. Showing a physical presence online is equally important to communicate teaching presence apart from ensuring that emails
are promptly answered and the timely return of assignments.

These individual and collaborative learning activities are embedded
under the cognitive and social presence and where all three elements
interact and are present, learning occurs and learners have a fruitful and
rich educational experience. It is therefore not surprising that the ways
university curricula are structured, the ways the content is created, delivered and received, and the ways assessments and evaluations are designed,
all go to frame what "digital technology" is for many university students
(Henderson et al., 2017, p. 1577) during the COVID pandemic period.
Technology adoption has accelerated developments in learning analytics
which have fast evolved from early attempts to gain a better understanding of student learning (Siemens & Long, 2011). However, many institutions still make surprisingly little use of assessment and marking analytics
beyond students' academic results (Ellis, 2013). Assessment is significant
as it constitutes evidence of student learning (or the lack thereof) and
rubrics are important in assessing criteria-based marking and providing
feedback in teaching and learning (Blumberg, 2013; Panadero et al., 2013).

12.8 COMMUNITY OF PRACTICE

Wenger et al. (2002) describe communities of practice (COP) as "groups
of people who share a concern, a set of problems or a passion about a topic
and who deepen their knowledge and expertise in this area by interacting
on an ongoing basis" (p. 4). A COP connects and collaborates individuals
with others to read, share, ask questions, and discuss a shared interest. It is
a kind of social learning in a participatory process (Tummons, 2018) that
can take place anytime, anywhere, face-to-face in online spaces or a combination of both and can involve people within and outside your discipline
and institution (Tutalo, 2019).

Online environments create opportunities for networking and collaboration that are flexible and are not limited by time and location constraints. COPs were formed to support the transition to a blended learning
environment and encourage best practice sharing, reflection and mutual
support. For example at JCU Singapore, we got past some of the challenges
of working from home and starting working on how we can assist faculty with assist with teaching strategies and resources and providing all
the necessary support via the JCU Singapore COP – which is made up
of an informal group of people coming together to share and learn from

one another in a manner that will facilitate and enhance their teaching practices and make them more aware of learner experience. There have been various COPs that came together to learn and use various online resources such as digital pen tablet, online whiteboard, interactive html-5 packages as well as sharing of a teaching design model during a crisis like the pandemic.

Besides, JCU Singapore also entered into a collaboration with a publisher to enhance student online experience through professionally developed and curated content that complements synchronous delivery in virtual and on-campus classes. This project involves the development of several subjects as a blended learning experience. These online modules are designed based on a collaborative planning effort, leveraging existing content and implemented within our Learning Management System (LearnJCU).

12.9 BENEFITS OF DIGITAL TECHNOLOGIES ON LEARNING AND TEACHING

The COVID-19 crisis has brought together faculty and staff from across the departments to find solutions for the evolving educational landscape. At JCU Singapore, faculty and staff have worked with specialists, vendors, developers and suppliers to provide resilient physical and digital infrastructure to suit the crisis conditions. For instance, the use of proprietary virtual classroom tool BlackBoard Collaborate™ across JCU took on a manifold increase since the pandemic started in early 2020. The use of educational technology was underpinned by hands-on support and an evolving set of just-in-time resources for both students and faculty (JCU, 2020c).

The benefits of digital technologies during the COVID-19 pandemic include an increase in blended and more accessible digital forms of education to support life-long learning (Beech & Anseel, 2020). There is increasingly a plethora of personalised, remote, adaptive and data-driven sources which enable the efficient delivery and personalisation of learning processes (Henderson et al., 2017). Students can now use their mobile devices to access and participate in the online environment (Gikas & Grant, 2013) thereby enhancing the diverse provision and equity of access to higher education.

For example JCU Singapore students could attend synchronous virtual classes from any location using any internet-connected device. In these sessions, students interact with their lecturers and fellow students using the chat function, breakout room function and/or the video in synchronous virtual classes via the Blackboard Collaborate™ available in the LearnJCU learning management system (LMS). They could also access pre-recorded content at a time convenient to them (asynchronous learning) and email the teaching staff should they have further queries. Such form of learning increases the convenience and flexibility of online education which can be very effective during the pandemic (McBrien et al., 2009; Schoech, 2000).

In the area of pedagogical practice research, the crisis has provided an impetus for COVID-19 research, with many researchers and teachers taking a problem-oriented approach seeking to address the challenges associated with the pandemic effect on learning and teaching. Researchers are conducting collaborative research with businesses and community stakeholders instead of about business and society (Beech & Anseel, 2020). For instance, two academics from JCU Singapore have developed a dynamic learning design model to support teaching readiness in uncertain times. Faced with urgency to shift teaching to an online platform due to the pandemic, a G-READY model (Sabharwal & Chawla, 2021) was conceptualised and developed. Another faculty member is collaborating with a group of researchers from Australia and New Zealand on a research project titled "Teaching & Learning in COVID-19 times" (Phillips et al., 2020) to investigate "the innovation, novel partnerships, and enhanced questions of access in teaching and learning practices due to COVID-19 precautionary measures" (JCU, 2020a).

12.10 CHALLENGES OF COVID-19 PANDEMIC ON LEARNING AND TEACHING

The pandemic has significantly changed the way we carry out our normal activities thereby impacting on our behaviours to a certain extent. This is not without significant difficulties and disruption. For instance, physical and social distancing requirements constrain activities on-campus and in public spaces. Online sessions are limited by technology and digitals skills. The ability to access and navigate digital platforms has become a cornerstone of success. While some take to online learning and teaching readily, others are confronted with an adoption issue in teaching, learning, and technological readiness that made problems of equity and access more apparent than ever (Buchholz et al., 2020).

UNESCO warns of the widening digital divide with an estimated two-thirds of all students worldwide still affected by disruption (Tawil, 2020). Meanwhile 95% of students in Switzerland, Norway and Austria have a computer to use for their schoolwork, only 34% in Indonesia do, according to OECD data. In the United States, there is a significant gap between those from privileged and disadvantaged backgrounds and while virtually all 15 year olds from a privileged background said they had a computer to work on, nearly 25% of those from disadvantaged backgrounds did not (Li & Lalani, 2020).

Moreover, the gap in digital know-how becomes apparent in the assumption that digital natives know how to use technology to navigate their learning and to conduct academic research. The area relating to professional conduct and integrity becomes highly relevant – how to do research and communicate in the academic and professional world. It is about responsibilities and how to prepare our graduates for the real world. The ability to learn, reassess, continuously adapt and expand our knowledge is the key literacy skill in an ever-evolving world of rapid change. The need for digital literacy and ability to adapt to the changing environment is further manifested through online educational experiences evolving as a form of crisis management thus triggering digital citizenship in the post-COVID-19 world (Buchholz et al., 2020).

Feedback from both staff and students highlighted the many challenges faced in this mode of learning and teaching. Key issues faced by learners and teachers in JCU Singapore include:

- Connectivity problems
- Awareness of technical requirements and browser settings
- Limited ability to use LearnJCU (LMS) portal for online learning and teaching to its full potential
- Access to online learning and teaching resources such as books, software and tools for research, writing and data analysis
- Awareness of conventions and best practice for self-directed online study
- Difficulty in finding suitable collaborators or project partners

Moreover, online interaction can be challenging due to the inability to read non-verbal cues such as facial expression, gestures and body language (Keating, 2020) even though some have used emoji and text representation in the chat boxes found online (Gajadhar & Green, 2005). There are also constraints in physical face-to-face gathering where group size is limited by social distancing, face masks are required and participants have to maintain physical distance. For example, close proximity interactions such as group work need to adapt to social distancing requirements and fieldwork is becoming more difficult due to restrictions in the number of participants or site access. To that extent, we are forced to rethink how to conduct these activities.

Access to common and shared facilities such as computer labs and the library was made difficult or near impossible due to safe distancing and hygiene considerations. This poses concerns about equity, access and accessibility and makes it difficult to conduct collaborative activities in face-to-face settings. The nature of these learning and developmental activities means that field trips, lab-work, student club activities, contact sports, performing arts and extra-curricular activities are difficult to stage virtually. At JCU Singapore, our students reported (ifocus groups, reflections) that teamwork has become more challenging due to difficulty in locating team members and sustaining collaboration overtime.

While student feedback shows that we can do more, adopting online teaching in a short time span requires our teaching staff to tackle big challenges one at a time to enhance the learner experience. Discipline specific problems included sharing of mathematical problems, equations, graphs and sketches. An example from a quantitative subject demonstrates the effectiveness of collaborative support networks and iterative processes to help to teach staff to adopt new technologies and build their abilities, skills and confidence. Teaching numeracy concepts with digital technologies is a challenge as what seems easy to do with pen and paper/whiteboard is not necessarily easily replicated online. While collaborative classrooms feature a virtual whiteboard, their capabilities are limited and do not easily support complex sketching, problem and equation solving. Our ICT staff worked tirelessly to address the problem of stable and secure access to the university's digital spaces. In collaboration with LearnJCU support and vendors, a VPN solution leveraging a proprietary Cloud service was established for learners in mainland China who are unable to return to Singapore because of restrictions to international arrivals.

Easing of social distancing restrictions for face-to-face in-person activities enhanced affordances to blended learning activities. Blended learning-enabled learners outside Singapore to participate and engage but such setups require careful planning and support for both staff and learners to manage the technological, administrative and social affordances of such environments (Bower et al., 2014; Li et al., 2020), some of which are beyond control, such as scheduling classes across time zones, internet bandwidth at client base or connectivity issues.

12.11 SKILLS AND CAPABILITIES FOR THE CHANGING ENVIRONMENT

The COVID-19 pandemic has brought a sharper focus to the types of skills and capabilities for digital transformation in higher education. In confronting an unprecedented situation of a lifetime, students and educators learn about management in a crisis. They discover and adopt the skills for survival – being adaptable, creative, innovative, resilient, agile, nimble and learning to pivot to new learning and teaching digital environment.

There is also the need to consider the types of capabilities to engage in the ABC of learning and teaching across a range of experiences. These include:

- Academic culture

- Business and industry engagement

- Community engagement

12.12 ACADEMIC CULTURE

The academic culture is about the ways we communicate, collaborate and share our values, knowledge and ideas within our institution and with external stakeholders to build and create new knowledge. Therefore it is essential to keep the communication going during times of lockdown and social distancing. In the area of learning and teaching, the paradoxical closure of the physical space and opening up of the digital space has created new opportunities for online learning and teaching.

The online mode encourages self-directed learning as students take charge of their learning, engage in learning individually and interact with the content, fellow students and the teacher (Anderson & Garrison, 1998; Brookfield, 2009; Garrison, 1997; Wenger, 1998) besides the industry experts and other professional staff at the institution. For example at JCU Singapore, the Learning Centre in collaboration with the campus activities department implemented "DigiLearn", a digital access and literacy programme for all new students that focuses on access to LearnJCU LMS and the ability to successfully use the online learning environment, online safety and integrity, communication and

234 ■ Digital Transformation in a Post-COVID World

problem solving when encountering issues on the digital platform. This programme serves to educate new students in the know-how and use of LearnJCU and build digital literacy skills for academic learning and research purposes.

Learning activities in higher education are not limited to the classroom and the mastering of digital skills. Life-long learning requires individuals to take control of their learning and constructively engage, communicate and collaborate with others whether in private life, at the workplace or in a social setting. To this effect, digital technology has become (and will continue to be) a key enabler for access to information, learning opportunities and social interaction at a local and global level (UNESCO, 2020). In the formal higher education context, an educator's role is one of a knowledge broker who facilitates discourse, learners' reflection and interactions with the course material and other students. Blended learning opens up new opportunities to extend interactions across time, place and space (Garrison & Vaughan, 2008).

The academic culture extends beyond the delivery of synchronous and asynchronous learning activities during this period of the pandemic. It is also about the whole of institution engagement with research in collaboration with the business and industry sectors. JCU Singapore is a key contributor to research into the tropical ecosystems, conservation and climate change; industries and economies in the tropics as well as people and societies in the tropics.

Professorial lectures, research forums, conferences and "brown/green bag" lunchtime seminars transitioned to online format to showcase research expertise across the disciplines and provide opportunities for knowledge sharing and co-creation, networking and discourse amongst students, faculty, external stakeholders and the broader society. Several online events reached out to an international audience and the new virtual format allowed for interaction between the presenter and audience beyond a short Q&A at the end of the session. Virtual events such as three-minute thesis (3MT) competition and graduation ceremonies continued to celebrate research and student achievements. Other events such as e-career fairs and virtual university open houses sought to maintain connection with staff, students, alumni and employers. Telecommuting and work from home guidelines and necessary resources were provided to assist researchers working remotely and minimising disruption to research activities. With the end of the Singapore circuit breaker, small group face-to-face research activities are now possible again.

12.13 BUSINESS AND INDUSTRY ENGAGEMENT

Business and industry engagement has become a key enabler driving learning innovation and social impact in higher education. We continued to maintaining and build relationships with strategic partners through research collaboration, research translation, knowledge transfer, internships and industry projects. While the move to virtual spaces has affected learning and teaching at its core, new opportunities have opened up allowing access to high-profile guest speakers in a virtual format at national and international level. Guest speakers from industry and webinars exposed students to business and management in diverse contexts and many of these sessions provided unique learning experiences since industry experts alike were forced to respond to rapid change and constant disruption and their learnings reflected in the talks.

Project-based learning across a range of subjects challenges students to tackle real-world problems and apply knowledge and skills learned in the classroom. Some of these learning experiences showcased in the Convergence Conference at JCU Singapore are co-facilitated with business organisations and other non-profit organisations such as the Food Bank Singapore. Members of the Student Marketing Club were selected into an Industrial Attachment programme with TAIGER which involved remotely conducted research amid social distancing measures (JCU, 2020b). Other student teams worked on real-world, industry-linked projects such as the application of statistical process control in chocolate manufacturing and reusing coffee powder waste for earthworm farming. Such multi-disciplinary projects challenged students to develop employability and problem-solving skills.

12.14 COMMUNITY ENGAGEMENT

The university's work contributes to solving pressing challenges in social justice, sustainability and equity through in-depth, impartial research and assessment. Global alliance of institutions such as the State of the Tropics aims to provide a solid foundation for decision-makers, policymakers, geopolitical analysts and other stakeholders. Community engagement also includes efforts such as the provision of grants for students in need, scholarships, corporate social responsibility events and engagement with the wider Singapore and ASEAN community. Throughout 2020, staff and students continued to engage in virtual and physical projects that make a tangible difference to life in the tropics such as workshops on equity issues in

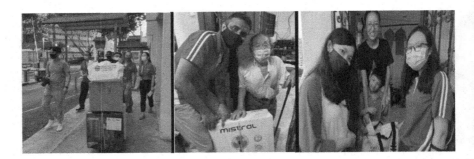

FIGURE 12.2 Students and staff distributing Christmas gifts to residents on 23 December 2020. (https://www.facebook.com/jcu.singapore.fanpage/photos/pcb.4020757874603265/4020727014606351/?type=3&theater)

tourism, beach clean-ups, tree-planting activities or the annual Christmas Community Project (Figure 12.2). The university liaises with diverse stakeholders through various educational activities, thought leadership and engagement. The mode of delivery in these events changed from in-person event to virtual event and social media such as Facebook, LinkedIn, Twitter and the JCU Singapore website. These platforms became the foundations for JCUS' thought leadership strategy and social networking channels.

12.15 IMPLICATIONS ON THE FUTURE OF LEARNING AND TEACHING IN HIGHER EDUCATION POST-COVID-19

A survey of recent literature suggests that our experiences align with those of other empirical research studies. Embracing change in a steady reflective manner can become the constant in an ever-changing environment (Ng, 2020). Research by Brammer & Clark (2000) and Lim (2020) highlight the need for consistent and clear communication at all levels, paired with agility and resourcefulness. Pather et al. (2020) who draw on reflections of anatomy teachers across Australia and New Zealand identify six critical elements for successful responses to online learning in response to the pandemic: "community care, clear communications, clarified expectations, constructive alignment, COP, ability to compromise and adapt and continuity planning" (p. 285).

The above examples illustrate the importance of a whole of the university experience for the growth and development of learners and educators alike, even more so during the crisis period with heightened apprehension and uncertainty surrounding their learning and teaching. From the learners' perspective, the ever-changing nature of technology highlights the need to continually reskill and acquires new digital competencies throughout their lifetime such that life-long learning becomes essential (Passey et al., 2018) to prepare them for the real business world – an environment that is often volatile, uncertain, complex and ambiguous. The world has changed to a large extent since the pandemic and so have the ways of working where these buzz words hold the keys to success: pivoting, adapting, resilience, nimble, creative and innovative.

On the part of educators and professional staff, COVID-19 pandemic highlights the capacity of JCU Singapore in adapting and pivoting to the changed circumstances by actively re-engineering the academic side of things (including changes to curriculum design, assessments, use of online tools in teaching and pursuing active research to track learner outcomes), rethinking the way we engage with the business, industry sector as well as the community. There is also the impetus to rethink student mobility, work-integrated learning and how we might best prepare our graduates for an uncertain future in a changing world of work (Kit & Co, 2020).

It is recognised that the current disruption leads to innovation in ways that engage learners in digital content and activities. It is also pertinent to consider the sum of the student experiences both online and face to face that create a sense of identity and belonging to the student body and institution of study that students may not have the opportunity to visit physically. Successful learning experiences are about how well we manage to connect productive inquiry on an individual basis and as a community (Anderson, 2018).

Having navigated the rapid transition to fully online learning and teaching setup, crisis management needs to shift into strategic planning mode quickly to rectify the gaps identified through reflection in the digital space. There is a need to stay mindful of the widening digital divide and the technical issues (many of which are beyond our control) that continue to disrupt learning and teaching online. Further work is required in regard to academic integrity in online assessment and examinations. In recent years academic integrity and contract cheating have become a growing concern (Awdry, 2020).

While technology such as plagiarism checks, monitoring software or introducing assessment settings such as time limits or randomness can mitigate to a certain extent, the root of the problem lies elsewhere. The shift to fully online learning requires us to investigate authentic assessment methods together with consideration for integrity and assurance of learning outcomes (JCU, 2020a). While online assessment and online marking were in use across all subjects already, some assessment tasks did not easily translate into applied examples and invigilated online exams posed a considerable challenge. Unless we manage to broadly integrate academic integrity into the culture, success is unlikely (Bretag et al., 2019). It takes more than just educating students about the concept of integrity. They need to understand what is acceptable, where and how to reach out for support and learn to use evidence-based claims for special considerations. There is also a need to equip themselves with information and digital literacy skills along with self-management skills to avoid academic misconduct.

12.16 CONCLUSION

In the age of digital media, Lefebvre's work (Lefebvre et al., 1991) reminds us that digital technology can create spaces of alienation and what is needed is a humane digital society with a self-managed, socialised Internet and digital media landscape where digital citizenship prevails (Fuchs, 2018).

More than that, it is about the connection with people and providing support for the well-being and mental health of both learners and educators. At JCU Singapore, the weekly JCUCARES: CONNECT informal sessions brought staff online to share tips on coping with working from home, staying safe and sane during the lockdown period. Workshops conducted by the Counselling Centre and the Clinical Psychology aim for work–life balance and mental well-being. Likewise, a series of online workshops were organised for students on mental health issues ranging from topics on stress of studying alone to coping and resilience.

This study suggests that the COI and COP frameworks are useful in highlighting the significance of social interactions and participation to bring about learning on the part of students and academics in the realm of the digital environment.

It is therefore pertinent to take on a learner-centric process that focuses on integrating new students into the University community by providing guidance and support during their initial adjustment to university life. This means that institutions need to think of different ways of orientating learners to the tertiary education experience and a holistic approach to the onboarding of students in the hybrid model of virtual with face-to-face experiences. A key driver of these changes is the learner experience and ability of everyone involved to effectively acquire skills and attributes for the new digital workplace.

REFERENCES

Akyol, Z., Garrison, D. R., & Ozden, M. Y. (2009). Online and blended communities of inquiry: Exploring the developmental and perceptional differences. *The International Review of Research in Open and Distributed Learning, 10*(6), 65–83. https://doi.org/10.19173/irrodl.v10i6.765.

Anderson, T. (2008). *The theory and practice of online learning* (2nd ed.). AU Press.

Anderson, T. (2018 February 2). How communities of inquiry drive teaching and learning in the digital age. *TEACHONLINE.CA*. https://teachonline.ca/tools-trends/how-communities-inquiry-drive-teaching-and-learning-digital-age.

Anderson, T., & Garrison, D. R. (1998). Learning in a networked world: New roles and responsibilties. In *Distance Learners in Higher Education: Institutional responses for quality outcomes*. Madison, WI: Atwood.

Awdry, R. (2020). *Contract cheating*. Teqsa. https://www.teqsa.gov.au/preventing-contract-cheating.

Beech, N., & Anseel, F. (2020). COVID-19 and its impact on management research and education: Threats, opportunities and a manifesto. *British Journal of Management, 31*(3), 447–449. https://doi.org/10.1111/1467-8551.12421.

Blumberg, P. (2013). *Assessing and improving your teaching: strategies and rubrics for faculty growth and student learning* (First ed.). Jossey-Bass.

Bower, M., Kennedy, G., Dalgarno, B., Lee, M. J., & Kenney, J. (2014). *Blended synchronous learning: A handbook for educators*. Office for Learning and Teaching, Department of Education.

Brammer, S., & Clark, T. (2020). COVID-19 and Management Education: Reflections on challenges, opportunities, and potential futures. *British Journal of Management, 31*(3), 453–456. https://doi.org/10.1111/1467-8551.12425.

Bretag, T., Harper, R., Burton, M., Ellis, C., Newton, P., van Haeringen, K., Saddiqui, S., & Rozenberg, P. (2019). Contract cheating and assessment design: Exploring the relationship. *Assessment & Evaluation in Higher Education, 44*(5), 676–691.

Brookfield, S. D. (2009). Self-Directed Learning. In R. Maclean & D. Wilson (Eds.), *International handbook of education for the changing world of work: Bridging academic and vocational learning* (pp. 2615–2627). The Netherlands: Springer. https://doi.org/10.1007/978-1-4020-5281-1_172.

Brown, C., & Salmi, J. (9 April 2020). Readying for the future: COVID-19, higher ed, and fairness. *Today's students tomorrow's talent*. https://medium.com/todays-students-tomorrow-s-talent/readying-for-the-future-covid-19-higher-ed-and-fairness-f7eeb814c0b8.

Buchholz, B. A., DeHart, J., & Moorman, G. (2020). Digital citizenship during a global pandemic: Moving beyond digital literacy. *Journal of Adolescent & Adult Literacy, 64*(1), 11–17. https://doi.org/10.1002/jaal.1076.

Chundur, S., & Prakash, S. (2009). *Synchronous vs asynchronous communications – What works best in an online environment? Lessons learnt*. EdMedia + Innovate Learning, Honolulu, HI, USA. https://www.learntechlib.org/p/31991.

Daniels, H. (2012). Institutional culture, social interaction and learning. *Learning, Culture and Social Interaction, 1*(1), 2–11. https://doi.org/10.1016/j.lcsi.2012.02.001.

Ellis, C. (2013). Broadening the scope and increasing the usefulness of learning analytics: The case for assessment analytics. *British Journal of Educational Technology, 44*(4), 662–664. http://eprints.hud.ac.uk/id/eprint/16829/.

Falloon, G. (2011). Making the connection: Moore's theory of transactional distance and its relevance to the use of a virtual classroom in postgraduate online teacher education. *Journal of Research on Technology in Education, 43*(3), 187–209. https://doi.org/10.1080/15391523.2011.10782569.

Fuchs, C. (2018). Towards a critical theory of communication with Georg Lukács and Lucien Goldmann. *Javnost – The Public, 25*(3), 265–281. doi: 10.1080/13183222.2018.1463032.

Gajadhar, J., & Green, J. (2005). The importance of nonverbal elements in online chat. *Educause Quarterly, 28*(4), 63.

Garrison, R. (1997). Self-directed learning: Toward a comprehensive model. *Adult Education Quarterly, 48*, 18–33. https://doi.org/10.1177/074171369704800103.

Garrison, D. R. (2007). Online community of inquiry review: Social, cognitive, and teaching presence issues. *Journal of Asynchronous Learning Networks, 11*(1), 61–72.

Garrison, D. R., Anderson, T., & Archer, W. (1999). Critical inquiry in a text-based environment: Computer conferencing in higher education. *The Internet and Higher Education, 2*(2–3), 87–105. https://doi.org/10.1016/s1096-7516(00)00016-6.

Garrison, D. R., & Cleveland-Innes, M. (2005). Facilitating cognitive presence in online learning: Interaction is not enough. *American Journal of Distance Education, 19*(3), 133–148. https://doi.org/10.1207/s15389286ajde1903_2.

Garrison, R., & Vaughan, N. D. (2008). *Blended learning in higher education: Framework, principles, and guidelines* (1st ed.). Jossey-Bass.

Gikas, J., & Grant, M. M. (2013). Mobile computing devices in higher education: Student perspectives on learning with cellphones, smartphones and social media. *The Internet and Higher Education, 19*, 18–26. https://doi.org/10.1016/j.iheduc.2013.06.002.

Henderson, M., Selwyn, N., & Aston, R. (2017). What works and why? Student perceptions of 'useful' digital technology in university teaching and learning. *Studies in Higher Education, 42*(8), 1567–1579. https://doi.org/10.1080/03075079.2015.1007946.

Hrastinski, S. (2008). Asynchronous and synchronous e-learning. *Educause Quarterly, 31*(4), 51–55. https://er.educause.edu/articles/2008/11/asynchronous-and-synchronous-elearning.

JCU. (2020a). *Designing Alternative Assessment*. https://www.jcu.edu.au/learning-and-teaching/assessment@jcu/designing-alternative-assessment.

JCU. (2020b). *In an age of work-from-home, students from the JCU Marketing Club succeed in online collaboration with Singapore AI start-up, TAIGER.* Members of the JCU Marketing Club were selected into an Industrial

Attachment program with TAIGER and conducted regular interactions and research, remotely, amid social distancing measures.

JCU. (2020c). *Teaching@JCU*. https://www.jcu.edu.au/learning-and-teaching/teaching@jcu.

Keating, E. (20.10.2020). Why do virtual meetings feel so weird? *Culture Lab*. https://www.sapiens.org/language/nonverbal-communication-online/.

Kit, T. S., & Co, C. (2020, October 17). IN FOCUS: Graduating into a COVID-19 jobs market – short-term challenges and longer-term issues? *CNA Channel News Asia*. https://www.channelnewsasia.com/news/singapore/in-focus-covid-19-graduates-singapore-jobs-market-employment-13168366.

Laurillard, D. (2002). *Rethinking university teaching: a conversational framework for the effective use of learning technologies* (2nd ed.). RoutledgeFalmer.

Lefebvre, H., Nicholson-Smith, D., & Harvey, D. (1991). *The production of space*. Oxford, UK: Blackwell Publishing.

Li, C., & Lalani, F. (2020 April 20). The COVID-19 pandemic has changed education forever. This is how. *The World Economic Forum COVID Action Platform*. https://www.weforum.org/agenda/2020/04/coronavirus-education-global-covid19-online-digital-learning/.

Li, X., Yang, Y., Chu, S. K. W., Zainuddin, Z., & Zhang, Y. (2020). Applying blended synchronous teaching and learning for flexible learning in higher education: An action research study at a university in Hong Kong. *Asia Pacific Journal of Education*, 1–17. https://doi.org/10.1080/02188791.2020.1766417.

Lim, M. (2020). Educating despite the COVID-19 outbreak: Lessons from Singapore. *Times Higher Education*. https://www.timeshighereducation.com/blog/educating-despite-covid-19-outbreak-lessons-singapore.

McBrien, J. L., Cheng, R., & Jones, P. (2009). Virtual spaces: Employing a synchronous online classroom to facilitate student engagement in online learning. *The International Review of Research in Open and Distributed Learning*, 10(3). https://doi.org/10.19173/irrodl.v10i3.605.

Moore, M. G., & Kearsley, G. (2011). *Distance education: A systems view of online learning*. Cengage Learning.

Ng, P. T. (2020). Timely change and timeless constants: COVID-19 and educational change in Singapore. *Educational Research for Policy and Practice*, 1–9.

Nordmann, E., Calder, C., Bishop, P., Irwin, A., & Comber, D. (2019). Turn up, tune in, don't drop out: The relationship between lecture attendance, use of lecture recordings, and achievement at different levels of study. *Higher Education*, 77(6), 1065–1084. doi:http://dx.doi.org.elibrary.jcu.edu.au/10.1007/s10734-018-0320-8.

O'Brien, D., & Scharber, C. (2008). Digital literacies go to school: Potholes and possibilities. *Journal of Adolescent & Adult Literacy*, 52(1), 66–68. https://doi.org/10.1598/JAAL.52.1.7.

O'Callaghan, F. V., Neumann, D. L., Jones, L., & Creed, P. A. (2017). The use of lecture recordings in higher education: A review of institutional, student, and lecturer issues. *Education and Information Technologies*, 22(1), 399–415. doi:http://dx.doi.org.elibrary.jcu.edu.au/10.1007/s10639-015-9451-z.

Panadero, E., Jonsson, A. (2013). The use of scoring rubrics for formative assessment purposes revisited: A review. *Educational Research Review, 9*, 129–144. https://doi.org/10.1016/j.edurev.2013.01.002.

Passey, D., Shonfeld, M., Appleby, L. et al. (2018). Digital agency: Empowering equity in and through education. *Tech Know Learn 23*, 425–439. https://doi.org/10.1007/s10758-018-9384-x.

Pather, N., Blyth, P., Chapman, J. A., Dayal, M. R., Flack, N. A. M. S., et al. (2020). Forced disruption of anatomy education in Australia and New Zealand: An acute response to the COVID-19 pandemic. *Anatomical Sciences Education, 13*(3), 284–300. https://doi.org/10.1002/ase.1968.

Phillips, L. G., Coleman, K., & Burke, G. (2020). *Teaching & learning in COVID-19 times study: an art & science collaboration*. JCU. https://www.jcu.edu.au/college-of-arts-society-and-education/postgraduate-study-and-research/education-research/research-projects/teaching-and-learning-in-covid-19-times.

Reidsema, C., Kavanagh, L., Hadgraft, R., & Smith, N. (2017). *The flipped classroom: Practice and practices in Higher Education*. Singapore: Springer.

Rourke, L., & Kanuka, H. (2009). Learning in communities of inquiry: A review of the literature. *International Journal of E-Learning & Distance Education, 23*(1), 19–48.

Roy, R., & Uekusa, S. (2020). Collaborative autoethnography: "self-reflection" as a timely alternative research approach during the global pandemic. *Qualitative Research Journal, 20*(4), 383–392. https://doi.org/10.1108/QRJ-06-2020-0054.

Sabharwal, J., & Chawla, S. (2021). *A G-READY model to support subject design for software engineering*. Conference: Joint Proceedings of SEED & NLPaSE co-located with 27th Asia Pacific Software Engineering Conference 2020. http://ceur-ws.org/Vol-2799/Paper2_SEED.pdf.

Schoech, D. (2000). Teaching over the internet: Results of one doctoral course. *Research on Social Work Practice, 10*(4), 467–486. https://doi.org/10.1177/104973150001000407.

Seixas, P. (1993). The community of inquiry as a basis for knowledge and learning: The case of history. *American Educational Research Journal, 30*(2), 305–324. https://doi.org/10.3102/00028312030002305.

Shah, D. (2019 December 17). Online degrees slowdown: A review of MOOC stats and trends in 2019. *The Report*. https://www.classcentral.com/report/moocs-stats-and-trends-2019/.

Siemens, G., & Long, P. (2011). Penetrating the fog: Analytics in learning and education. *EDUCAUSE Review, 46*(5), 30.

Swan, K., Garrison, D. R., & Richardson, J. C. (2009). A constructivist approach to online learning: The community of inquiry framework. In R. P. Carla (Ed.), *Information technology and constructivism in higher education: Progressive learning frameworks* (pp. 43–57). IGI Global. https://doi.org/10.4018/978-1-60566-654-9.ch004.

Tan, H. R., Chng, W. H., Chonardo, C., Ng, M. T., Fung, F. M. (2020). How chemists achieve active learning online during the COVID-19 pandemic: Using the community of inquiry (CoI) framework to support remote teaching. *Journal of Chemistry Education, 97*(9), 2512–2518. https://doi.org/10.1021/acs.jchemed.0c00541.

Tawil, S. (2020). *Six months into a crisis: reflections on international efforts to harness technology to maintain the continuity of learning,* @UNESCO. https://unesdoc.unesco.org/ark:/48223/pf0000374561?locale=en.

Tummons, J. (2018). *Learning architectures in higher education: Beyond communities of practice.* Bloomsbury Publishing. https://www.bloomsbury.com/au/learning-architectures-in-higher-education-9781350130975/.

Tutalo, A. (2019). Day 1 – What is a community of practice. *ANU Coffe Courses.* https://anuonline.weblogs.anu.edu.au/2019/08/19/day-1-what-is-a-community-of-practice/.

UNESCO. (2020). *Embracing a culture of lifelong learning: contribution to the futures of education initiative* (978-92-820-1239-0). UNESCO Institute for Lifelong Learning. https://unesdoc.unesco.org/ark:/48223/pf0000374112.

von Glasersfeld, E. (1989). Cognition, construction of knowledge, and teaching. *Synthese, 80*(1), 121–140. https://doi.org/10.1007/BF00869951.

Watts, J. (2017). Beyond flexibility and convenience: Using the community of inquiry framework to assess the value of online graduate education in technical and professional communication. *Journal of Business and Technical Communication, 31*(4), 481–519. https://doi.org/10.1177/1050651917713251.

Wenger, E. (1998). *Communities of practice: Learning, meaning, and identity.* Cambridge University Press.

Wenger, E., McDermott, R., Snyder, W. M., & Books24x, I. (2002). *Cultivating communities of practice: A guide to managing knowledge* (1 ed.). Harvard business school press.

CHAPTER 13

Bolstering Biometric Data

Can Privacy Survive in a Post-COVID World?

Melissa Wingard

Phillips Ormonde Fitzpatrick, Melbourne, Australia

CONTENTS

DOI: 10.1201/9781003148715-13

13.1 2021

COVID-19 has dominated news headlines and for the majority of us, our lives since early 2020. It seemingly came out of nowhere but authorities seemed confident in most parts of the world that a relatively short lockdown would help slow the spread while a vaccine could be developed. For many living affected countries that chose this course of action, we have spent a significant amount of time in 2020 and now 2021 continuing to be affected by various forms of quarantine, curfews, lockdowns, restrictions on movements and social interactions, and 'circuit breaker' actions. It is becoming apparent that the pandemic has not dissipated in the manner that many expected or hoped that it would. Public safety has become the overriding consideration in every aspect of life.

To limit the spread of the disease and meet new public health density requirements, many have moved to hybrid working environments where for some working from home has become the 'new normal' and others work from home some days and in the office other days. This new approach to our day-to-day lives has led to a fundamental shift in the way that people consider how they want to work, where they want to live and what is important to them.

Collectively we have created a new language and terminology to accompany this pandemic, such as social distancing, contact tracing, variants, super spreader events, zoom fatigue, ISO, doomscrolling and quarantine. Even the World Health Organization created a new name, COVID-19, to distinguish this particular coronavirus from other coronaviruses. As part of this new world and the new language we are also learning a new manner of engaging and participating in our society. At every turn we face a crossroad: Do we disclose personal information to be able to continue to participate in our society in the same manner that we did before COVID-19? As we move out lockdowns which were implemented in response to the immediate threat and work towards a way of living with COVID-19, we face a new world where technology is essential and much of that technology is reliant on the collection and use of our personal information and biometric data.

In the book *Sophie's World*, Jostein Gaarder when trying to explain philosophy and uses the analogy of the universe being like a white rabbit being pulled out of a magician's hat. When we are born we exist on the outer most tip of the rabbit's fur looking into the magicians' eyes wondering at the impossibility of this amazing trick but as we grow to be adults we nestle down into the fur where it is safe and familiar (Gaarder, 1994). Only philosophers risk climbing to the perilous edge of the rabbit's fur to stare into the magician's eyes. The rest of society gets comfortable, used to living in deep in the rabbit's fur, and fails to question our very existence accepting the status quo for what it is. It seems that privacy is facing a similar fate, for a while, people began to stare into the eyes of the Silicon Valley magicians and ask them directly what information do they have on us, what information will they collect, how are they planning on using and storing our personal information, especially our biometric data. Which led to new and more powerful legislation being implemented or amended by governments to address the concerns of their citizens, including but not limited to the European Union's General Data Protection Regulation and the California Consumer Privacy Act. Even jurisdictions like the Republic of India and the People's Republic of China which have historically been reluctant to recognise an individual's right to privacy have begun to review and implement new privacy laws to ensure their place as a part of the global economy. As COVID-19 wears on and we face a fundamentally different world order where public health trumps all other human rights, will people bury themselves in the rabbit's fur? Accepting that the very cost of living in this post-COVID-19 society means abandoning privacy for the benefit of public health?

13.2 WHAT DO WE MEAN BY PRIVACY?

Before moving on in the consideration of what privacy looks like in a world where technology is paramount we first need to consider what is meant by 'privacy'. It is a term that is so often used but without a settled universal meaning or common understanding. To some it is a fundamental right, others it is a legal construct created by governments, or for others a preferential way of living (Nissenbaum, 2010). The terms data protection, data privacy, privacy are all used interchangeably and in a somewhat inconsistent manner (O'Brien, 2020).

The United Nations has sought to enshrine privacy as a fundamental human right in the Universal Declaration of Human Rights and Article 17 of the International Covenant on Civil and Political Rights. In the European Union, privacy is a right protected under Article 8 of the European Convention of Human Rights along with the Convention for the Protection of Individuals with regard to Automatic Processing of Personal Data (Council of Europe, 1981) and the Charter of Fundamental Rights. This helps to explain why the European Union has taken a stronger stance in relation to its laws for the protection of privacy. These conventions and treaties were the precursors to the 1995 Data Protection Directive and the 2016 General Data Protection Regulation often considered the 'gold standard' of privacy laws. Whilst many countries have taken inspiration from the General Data Protection Regulation or see an economic benefit in being adequate for European businesses transferring data, not all countries have or will adopt the same approach to privacy protection. In fact, according to the United Nations Conference on Trade and Development only 128 of 194 countries have legislation in place to deal with the security and protection of data and privacy. The application of those laws is widely varied (Data Protection and Privacy Legislation Worldwide | UNCTAD, n.d.).

Regardless of semantics or legislation, privacy is the desire by individuals to keep control of our personal information. We want the right to determine what personal information we share with whom, for what reason and how that personal information can be used and shared.

For this chapter, we will consider privacy and personal information by reference to the definition as set out in Article 4(1) of the GDPR whereby:

> 'personal data' means any information relating to an identified or identifiable natural person ('data subject'); an identifiable natural person is one who can be identified, directly or indirectly, in particular by reference to an identifier such as a name, an identification number, location data, an online identifier or to one or more factors specific to the physical, physiological, genetic, mental, economic, cultural or social identity of that natural person;

This definition of personal information or personal data is the lynchpin that holds together the concept of privacy, the information or data you hold must be such that a person is identified i.e. your name, address, date of birth or email address are capable of being identified i.e. an individual who lives in Singapore and who owns a 1962 Ferrari 250 GTO. Given the rarity of such a car, only a very small number of them were made by Ferrari to meet racing rules at the time, it is likely that with these two pieces of information one could identify the individual in question and certainly, a computer with access to many databases could identify the individual.

Personal information should not be confused with publicly available information, the fact that your name and email address may be made publicly available through the use of an internet search engine, does not remove the fact that any company or organisation that collects, holds, accesses, stores or uses that information is required to comply with the relevant local privacy laws. It is still information that is personal to you, identifies you and for which you are entitled to control the usage of such.

13.3 HOW HAS COVID-19 CHANGED LIFE WITH RESPECT TO THE COLLECTION OF PERSONAL INFORMATION?

13.3.1 Society

Coming out of the various forms of lockdowns that have been imposed to manage and contain COVID-19 outbreaks, public safety experts are promoting aggressive and sustained contact tracing efforts. Contact tracing is crucial to interrupt chains of transmission of COVID-19 in the community, the process involves gathering information from a person who tests positive about where they have been and who they have been in contact with during the time they were considered infectious. This enables public health authorities to isolate and quarantine others who have possibly been infected to prevent the further spread of COVID-19 in the community. Aggressive use of contact tracing, along with associated isolation and quarantining of possible infections, has been successful in many countries including Australia, New Zealand, Japan, Singapore and South Korea.

The data source used by contact tracing includes:

- Various forms of mobile 'COVID' tracking and tracing apps;
- The use of facial recognition to identify those that have been in the same vicinity as an active case;
- The use of geolocation data from smartphones;
- Heat maps generated using large data sets to be able to understand population movements;
- Using banking records to show the use of electronic methods of payment;
- The requirement to 'check-in' using a quick response (QR) code (a two-dimensional barcode that stores significant amounts of data when visiting various venues.

No longer can we make an anonymous visit to our local supermarket, pay by cash and return home untracked and untraced.

Contact tracing requires by its very nature vast amounts of personal information which is encouraging many, to look to leveraging biometric data to assist with the efficacy of such contact tracing efforts. Some of these tracking and tracing apps use biometric data to provide greater certainty to the government, for example in Poland the COVID tracing app uses biometric data to ensure that those persons who are infected with COVID-19 remain in quarantine. China has implemented wide-scale use of facial recognition technology to prevent those who have been infected from travelling around the country. Russia is utilising facial recognition technology to identify and track those persons who fail to meet their mandatory quarantine requirements (OECD.org, 2020). In many cases, the use of such biometric data is being conducted where there is a lack of specific guidance given, or explicit consent sought, from individuals as to the collection, use, disclosure, storage and deletion of biometric data.

The increased use of masks by the general public and the requirement to wear masks in many places has meant that existing technology, such as facial recognition is facing challenges in recognising individuals. To compensate for this obstruction, facial recognition technology is looking to use additional forms of biometric data to facilitate that identification. Mannerisms such as a person's gait or facial geometry are required as additional reference points for the facial recognition systems to accurately identify or to even have the potential to accurately detect and identify individuals.

Now as the vaccine is rolled out in many countries, the question turns to how those vaccinated are to be identified which is where biometric data and the use of vaccine passports are proposed. These controls are based on the biometric passport which many countries currently use. The consequences of this are unknown at present.

13.3.2 Our Working Life

By now, for those of us who can, we have been asked to work from home. This meant quickly pivoting from working in an office with a defined space and usually a defined time, to working from the dining room table, a hastily erected desk in the bedroom or for those that are lucky, dedicated home office. That has come with the need for companies and businesses to equip their workers with technological solutions and equipment that they need to work from home. In most cases, businesses have been able to successfully continue operating with a large part, or all, of their workforce, working from home. Whilst a technological marvel that so many were equipped with the ability to work from home so quickly, this has presented new challenges and differing situation challenges when it comes to privacy, data security and cybersecurity risks. These risks arise concerning the personal information needed by organisations about their employees to facilitate secure access to networks and environments, both physical and cyber in addition to the personal information collected and held by organisations about their clients. Previously it was possible to limit physical access to personal information, whereas now organisations are relying on their staff to manage and maintain physical access to, and to protect, the personal information that the organisation stores.

When contacting any organisation it has become increasingly likely the person you are engaging with is working from home. Whilst the service provided and the outcome of any contact may be the same as in-person, practically any personal information disclosed on that call is potentially at risk of being overheard or seen by other members of that household. Such disclosure is likely a breach of any privacy policy but in the circumstances, the options are limited where there is no option to speak face to face in a physical building. Being limited to transacting over the phone or via email has necessitated that additional personal information is shared so organisations can verify your identification and to meet their 'know your client' obligations. It is much harder to verify an individual when you are unable to see their identification and their face in person. This situation has arisen because governments have made decisions to address the immediate public safety concern.

Further to the collection of additional personal information by organisations to be able to verify their client's identity in a virtual world, we are also seeing an increased use of biometric data by those organisations to facilitate employees working in a physical office or to authenticate and identify employees working virtually. A 'COVID safe' office environment utilises biometric data to enable contact-less and touchless technology limiting the spread of the virus. That can mean using facial recognition rather than swiping a pass to access the office or voice recognition to direct elevators. To facilitate the use of this technology employers are, and will be, collecting greater amounts of personal information and biometric data on their employees.

13.3.3 Transactions

One word that was repeatedly used in 2020 was 'pivot', businesses were required to pivot into alternate sales and distribution channels or to different aspects of the business to continue operating in the face of widespread and in some instances, long-lasting lockdowns. A key component of many businesses pivoting strategy was to begin to sell their services and goods online. Take for example the hospitality industry, faced with closures or social distancing requirements that limited patron numbers, restaurants moved to home delivery for a variety of offerings including takeaway food, ready-to-eat meals, online grocery or wine stores. Bars and pubs moved to selling any stock they had online either via webpages or using social media. Butchers, bakers, supermarkets and grocery stores all went online if they weren't already and began to offer home delivery. Online transactions necessitated the provision of additional amounts of personal information. At the very least, name, phone number and address are required to be able to make use of these new distribution channels. It raises the question as to whether these businesses were equipped or supported to be able to deal with the personal information collected, given the immediate need for alternate income streams.

Even as we emerge from our various lockdowns, either for a fleeting moment or in a more sustained release many businesses who retain a physical shopfront have moved from accepting cash to only accepting the card or touchless payment, i.e., mobile banking apps. This removes an individual's ability to engage in day-to-day transactions anonymously or to limit and control the spread and use of their personal information.

Suddenly we were faced with a world where even to obtain the basics of life we are required to share our personal information and our biometric data where the non-contact payment utilises smartphone technology protected by fingerprints as a means of authentication.

The risk that individuals face in the post-COVID-19 world is being denied a right to continue to participate in society if they are not willing to disclose personal information. At any other time, this would seem draconian, an infringement of civil liberties and we would be outraged. Whilst navigating this existence with COVID-19 concern for privacy and the protection of our personal information is being subjugated to a wider societal concern of catching a potentially fatal virus. Individuals may believe that they are equipped to weigh up the risks against the benefits but that assumes that we are being provided with the information necessary to make such a decision.

13.4 WHAT IS BIOMETRIC DATA AND WHAT DO THE PRIVACY LAWS PROVIDE?

Biometric data is personal information. Biometric data can be many things, but it is most commonly understood to mean fingerprints and facial recognition. Biometric data refers to the biological and physical characteristics used to identify someone including, but not limited to, iris scans, gait, ear shape, keystrokes, DNA, hand geometry, odour, voice recognition, speech recognition, vein recognition, body temperature and signature. All of these personal identifiers are classed as biometric data. Biometric data is unique in its ability to identify a person for their entire lifetime.

Article 4(14) of the General Data Protection Regulation defines biometric data as follows:

> 'biometric data' means personal data resulting from specific technical processing relating to the physical, physiological or behavioural characteristics of a natural person, which allow or confirm the unique identification of that natural person, such as facial images or dactyloscopic data;

The use of biometric data can facilitate discrimination, profiling and surveillance on a mass scale, which is why there must be strong legal frameworks, accompanied by strict controls and safeguards to manage the benefits of the collection of biometric data with the privacy and security risks faced by the individuals who own that biometric data. This is where the law has not been able to keep up with technology.

Increasingly technology providers are looking to utilising these various forms of biometric data in their technology to provide touchless or contactless technology for identity and access management, along with the collection and use as part of a concerted public health response to COVID-19 (Carlaw, 2020; Griffin, 2020).

We note that there is a distinction between using biometric data for identification and its use for authentication as part of cybersecurity protocols, for which there is a significant amount of debate and conjecture. This chapter does not propose to deal with those issues and for the purpose of this chapter, collection or use of the biometric data could be for either identification or authorisation. Privacy does not view the use of biometric data for identification purposes different from using the biometric data for authentication as part of an information security control. In both instances, the biometric data needs to be collected, used, stored, accessed and disclosed.

The critical and most crucial thing about biometric data is that it is inherently us. We generally have five fingers, two ears and one face. We cannot suddenly recreate a new nose like we can create a new password or learn how to walk differently easily. Even though we may provide the data willingly, and we remain as the owners of that data, often we lose control of the information and may give it in circumstances where we do not fully understand the consequences of providing such information.

Biometric data is considered a special category of personal information because of its very nature as being inherently and innately identifiable. Many of the privacy laws rely on the concept of anonymisation to enable data processors to utilise data for their interests or research without the need for ongoing compliance with privacy laws. The rationale for this approach is that once the personal information is de-identified or anonymised it is not possible to identify the individual, directly or indirectly, from that data. The ability to identify an individual from the data is the critical element of what comprises personal information. The data or information must be such that it is capable of identifying a person. This requirement for identification enables organisations with the ability to deal with personal information in any manner that they choose, provided it has been anonymised and the individuals are no longer able to be identified. The challenge with biometric data is that stripping the biometric data of its markers doesn't render it incapable of identifying someone. With the increase in computing power, smart algorithms and artificial intelligence it is possible for researchers using this technology to reorganise and reidentifies the biometric data such that it returns to its ability to identify individuals (Banda, 2019). Privacy professionals are increasingly expressing concern that biometric data can never truly be de-identified or anonymised. The current legislative stance taken by those countries with specific laws on the use of biometric data should be reconsidered in the absence of a true ability to anonymise biometric data.

There are a few laws around the world that do specifically address the use of biometric data most jurisdictions lack any laws specifically relating to biometric data, despite the increasingly extensive and invasive use of such data. According to Comparitech's biometric data study, the United States of America is ranked as one of the worst countries for the protection of biometric data, given that there are only a handful of states that have laws concerning the collection and use of biometric data, such as Illinois, and there is no specific federal law despite widespread use of facial recognition, fingerprints and biometric databases (Bischoff, 2020).

What Comparitech has seen in their studies is that countries that have poor protection for biometric data are also the countries that are implementing or looking to implement additional biometric controls to deal with the effects of the pandemic for example, the Peoples Republic of China use of drones with facial recognition technology to monitor residents as they move about their day to day lives, and mobile apps used to measure and record an individual's body temperature before boarding public transport (Bischoff, 2020).

The United States of America is looking at installing fever detection cameras, having facial recognition technology that can work even when persons are wearing masks. Amazon has launched Amazon One which is a contactless payment system that enables individuals to make payments in certain Amazon stores by hovering their palm over the Amazon One device, provided that your palm print and credit card details have been registered (Kumar 2020). The use of biometric data in our day to day lives is being used by both private organisations to provide 'COVID safe' services and by governments to manage and control a virus amid a pandemic.

Some jurisdictions do have laws specifically pertaining to the use, collection, storage and access to biometric data including the General Data Protection Regulation which applies across all member countries of the European Union. In most cases, though biometric data is captured under the definition of personal information such as in Singapore's Personal Data Protection Act (n.d.).

Article 9(1) of the General Data Protection Regulation prohibits the processing of biometric data, subject to the exceptions set out in the remainder of Article 9. These exceptions include where the data subject has given explicit consent for the processing of the biometric data, and where the processing of the biometric data "is necessary for reasons of public interest in the area of public health", as provided for in Article 9(2)(i).

In the United States of America, there is no single comprehensive federal law to address privacy but there are various state laws. Specifically, in respect of biometric data, Illinois has the Biometric Information Privacy Act passed in 2008 because it was concerned about the increased use and collection of biometric data by organisations (The Illinois Biometric Information Privacy Act ('BIPA'): When Will Companies Heed the Warning Signs?, 2020). The Biometric Information Privacy Act regulates how private entities collect, use and share biometric data. The California Consumer Privacy Act has also sought to regulate biometric data, although it does it through including a very detailed definition of biometric data as part of the definition of personal information.

In Australia, the *Privacy Act 1988 (n.d.)(Cth)* (**Privacy Act**) provides that biometric data is sensitive information and is therefore subject to more stringent requirements as to collection, use and storage of that data. However, there is an exception in circumstances where the collection, use or disclosure of the biometric data is necessary to lessen or prevent a serious threat to public health and safety. This is rather broad exception that enables organisations to collect, use and disclose biometric data for any purposes if it is necessary to lessen or prevent a threat to public health. The Australian Privacy Act is limited in its application however, it only applies to organisations with a turnover of more than $3 million, unless they are handling health information, are in the business of buying and selling personal information or have voluntarily opted in. Further, the Australian Privacy Act has an exception for personal information that employers hold on their current employees, which opens the door to the business's being able to use technology that uses biometric data to manage the workplace risks of COVID-19 without the need to adhere to any privacy controls.

Informed and freely given consent is a central tenant in privacy laws providing organisations to collect and use personal information, as we see below with the concept of the privacy paradox, consent is often given without full information on the possible uses of the data or where only the benefits and not the risks have been considered.

13.5 THE PRIVACY PARADOX AND COVID-19: HARDER FOR INDIVIDUALS TO ENGAGE IN PROTECTIVE ACTIONS

When surveyed individuals claim to care greatly about privacy but their actions rarely align with the concern expressed. That is whilst individuals generally express concern about the disclosure and use of their personal information when asked, but when confronted with the opportunity to take protective actions to secure their personal information online, they fail to do so. These protective actions could be limiting the amount of information disclosed, or else selecting not to use the online tools or service if it is requiring the disclosure of a disproportionate amount of personal information. This disconnect is referred to as the 'privacy paradox', a concept that was espoused by Susan Barnes in 2006 (Barnes, 2006). The privacy paradox has been explained in different ways including by reference to individuals finding it difficult to assign a specific value to their information or identify their information as their own, or that their decision making is flawed.

Over time the privacy paradox has operated differently as between online social networks such as Facebook and Instagram and mobile applications, including contact tracing or COVID safe apps. It was generally considered that when assessing whether one should provide personal information in exchange for the benefits of being able to participate online that the decision either consisted of a risk-benefit calculation or else the individual looked to the benefit alone (Barth, 2017).

The effect of the privacy paradox is amplified when it comes to mobile applications largely to time constraints and the need for immediate gratification (Barth & de Jong, 2017). The use of 'check in' or QR codes for venues to enable contact tracing emphasises the privacy paradox. Individuals may express concern about privacy, but when faced with the opportunity to see family or friends, most people will download any necessary app and input their personal information and agree to CCTV surveillance. The risks of sharing personal information or our biometric data are overridden by the situational benefits. It is difficult for individuals to fully assess the risks of such action as against an overwhelming desire to be a part of society. Failure to agree to these new conditions could mean that we are unable to socialise with our family and friends. An unthinkable prospect for those who have spent a large part of 2020 unable to see family and friends due to various border restrictions, lockdowns or movement limitations.

The challenge with the privacy paradox is that it presupposes that individuals have the knowledge and skills necessary to assess fully the risks associated with the disclosure of personal information, including biometric information (Barth & de Jong, 2017; Hoffman et al., 2016). However, in this world of concealed data practices and technological advances which permit smartphones to collect more information on individuals than is logically conceivable, it is not always possible for individuals to be able to fully assess the risk as against the rewards. Many individuals do not have the skills necessary to be able to carefully weigh the benefits of any online interaction or downloading of an app. Often we will see convenience and personalisation as benefits from the disclosure of personal information, without an understanding the information being disclosed isn't limited to name, email address or fingerprint. Geoffrey Fowler of the Washington Post in 2019 looked at the volume of information that our phones were giving out to apps in a week. He found that despite Apple's assertion that "what happens on your iPhone stays on your iPhone" his iPhone was providing a multitude of data trackers with various pieces of information from his phone, some he consented to, others he did not (Fowler, 2019). Mr. Fowler is an experienced technology writer who was unaware the extent to which apps on his phone were sharing information with third parties. To write the article he needed the assistance of an expert in this area to determine the full extent of the personal information and data that was being collected and shared by his phone. The average person is consistently going to be challenged to make a fully informed decision as to whether the benefit received from the disclosure of personal information is worth the risk. If we don't know where our data is going, who it is going to and for what reason how can we, as individuals even hope to protect it and to keep it private.

There is a belief held by many that the disclosure of their personal information is not that important because they are not that interesting. In many years of giving talks, presentations or advice on privacy, almost universally individuals say that they do not bother reading the privacy policy because there is not anything interesting about them, they are not doing anything wrong or they have already shared so much information why does it matter if they share anymore. This apathy is concerning as we move towards a world where biometric data is going to be the key data that underpins public health responses to COVID-19 and there is a lack of specific privacy protection.

Before COVID-19, with the technological advances and monopolistic behaviour of 'tech giants' the world had finally started to turn its mind to the importance of controlling our personal information, the European Union's General Data Protection Regulation was seen to be setting a world standard based on enabling the owner of the personal information to have greater control and rights with respect to our personal information (William, 2020). The California Consumer Protection Act comes into effect specifically targeted at the collection, use and exploitation of personal information which had become the core business of many technology companies in Silicon Valley.

COVID-19 has struck as the world was beginning to rethink privacy laws and encourage the promotion of privacy as a fundamental human right. The pandemic has shifted the focus to the public health actions necessary to limit the spread and destruction of COVID-19, including the use of privacy-invasive technology using biometric data. The Australian Human Rights Commission in its advisory report following the review of identity matching services legislation noted "the intrusions into citizens' privacy that are enabled by facial recognition technology are real, and they are profound. It has been said that technology may herald 'the end of anonymity'. For that reason, particular care needs to be taken to ensure that the use of biometric technologies including facial recognition technologies is strictly controlled" (Commonwealth of Australia, 2019).

13.6 WILL PRIVACY SURVIVE?

Privacy doesn't operate in isolation; COVID-19 has shown its intersection with public health, national security and economic development, specifically for governments. Where privacy legislation is in place, it will generally have exceptions and carve-outs on the use of personal information by governments for these purposes or else exempt the government entirely. As governments continue to fight COVID-19 it is not unreasonable to think that they will do so with all information that they have at their disposal. Privacy may become collateral damage in the war against the virus.

From an individual perspective, the perception of privacy and desire to enforce those rights may depend on your privilege and place in society. Will those employed in the gig economy or else starting in graduate positions feel as though they can say no if their employer requires them to submit to the use of biometric technology for them to continue to do their job? Will individuals resist technology that allows them to have dinner with friends because it's invading their privacy? To enforce their right to privacy for individuals, need a framework of laws that empower and enable us to control and manage the collection, use, disclosure and destruction of our biometric data alongside strong government intervention.

The longer that the pandemic continues and more the individuals get used to having their personal information and biometric data collected and used in the name of public safety and public health, the greater the likelihood that this will be accepted as part of the norm required to restore our society to the so-called new normal (Anderson et al., 2021).

The use of facial recognition has an even wider-ranging impact, not just on the question if privacy and the creation of a surveillance regime, but also on the question of guilt and innocence, the very basis of many western societies' judicial system. If a computer 'recognises you' as the 'guilty' party you have to prove your innocence, whereas the justice system usually operates on the basis that you are innocent until proven guilty, with often the provision of defences available to defend such conduct. If facial recognition identifies an individual having breached mandatory quarantine, the onus falls on the individual to prove their innocence. This is a fundamental shift in the way that western legal systems operate and the adoption of such will have long-reaching consequences for the justice system.

The pandemic continues. The vaccine that was supposed to save us is possibly flawed against the new variants that are emerging. Around the world individuals are dealing with mental health issues arising from prolonged or sustained lockdowns, money issues arising from unemployment or illness, fears of being infected with COVID-19, grief from watching others around them succumb to the virus. It should not fall to individuals alone to protect their privacy, especially not in this environment. It must be and needs to be a joint effort with governments and the technology companies that are using the biometric data. Those who can encourage the creation of technology that does not make individuals the product, that can address public health issues without sacrificing privacy. The very real risk is that privacy will not survive COVID-19. Individuals will grow accustomed to personal information and biometric data being harvested, collected, used, and stored to facilitate our place in society, to enable us to work and live. There is little benefit for those in power to assist individuals to exert their control and influence over their biometric data, as to do so may negatively impact any response to the immediate public health crisis being faced.

We won't know what we have given up until it's too late.

REFERENCES

Anderson, J., Rainie, L., & Vogels, E. A. (2021, February 18). Experts Say the 'New Normal' in 2025 Will Be Far More Tech-Driven, Presenting More Big Challenges. Pew Research Center: Internet, Science & Tech. https://www.pewresearch.org/internet/2021/02/18/experts-say-the-new-normal-in-2025-will-be-far-more-tech-driven-presenting-more-big-challenges/.

Banda, Justin (2019) *Inherently Identifiable: Is it possible to anonymise health and genetic data?* International Association of Privacy Professionals https://iapp.org/news/a/inherently-identifiable-is-it-possible-to-anonymize-health-and-genetic-data.

Barnes, S (2006) A privacy paradox: social networking in the United States. First Monday, 11(9) https://doi.org/10.5210/fm.v11i9.1394.

Barth, S., & de Jong, M. D. T. (2017). The privacy paradox – Investigating discrepancies between expressed privacy concerns and actual online behavior – A systematic literature review. Telematics and Informatics, 34(7), 1038–1058. https://doi.org/10.1016/j.tele.2017.04.013.

Bischoff, P. (2021, January 27). Biometric data: 96 countries ranked by how they're collecting it and what they're doing with it. Comparitech. https://www.comparitech.com/blog/vpn-privacy/biometric-data-study/.

California Consumer Privacy Act (CCPA). (2021, February 2). State of California - Department of Justice - Office of the Attorney General. https://oag.ca.gov/privacy/ccpa.

Carlaw, S. (2020). Impact on biometrics of Covid-19. Biometric Technology Today, 2020(4), 8–9. https://doi.org/10.1016/s0969-4765(20)30050-3.

Council of Europe. (1981). Convention 108 and Protocols. Data Protection. https://www.coe.int/en/web/data-protection/convention108-and-protocol.

Data Protection and Privacy Legislation Worldwide | UNCTAD. (n.d.). United Nations Conference on Trade and Development. https://unctad.org/page/data-protection-and-privacy-legislation-worldwide.

Fowler, G (2019). It's the middle of the night, do you know who your iphone is talking to? Washington Post https://www.washingtonpost.com/technology/2019/05/28/its-middle-night-do-you-know-who-your-iphone-is-talking/.

Gaarder, J. (1994). Sophie's World: A Novel about the History of Philosophy (1st ed.). Farrar, Straus and Giroux..

General Data Protection Regulation (GDPR) 2016 – Official Legal Text. (Regulation (EU) 2016/679 General Data Protection Regulation 2019, September 2). General Data Protection Regulation (GDPR). https://gdpr-info.eu/.

Griffin, J. (2020) *The Role of Biometrics in a Post COVID-19 World.* Security Info Watch. https://www.securityinfowatch.com/access-identity/biometrics/article/21143152/the-role-of-biometrics-in-a-post-covid19-world.

Hoffman, C. P., Lutz, C., & Ranzini, G. (2016). Privacy cynicism: A new approach to the privacy paradox. Cyberpsychology: Journal of Psychosocial Research on Cyberspace, 10(4), 7. https://doi.org/10.5817/cp2016-4-7.

Kumar, D. (2020, September 29). Introducing Amazon One—a new innovation to make everyday activities effortless. About Amazon. https://www.aboutamazon.com/news/innovation-at-amazon/introducing-amazon-one-a-new-innovation-to-make-everyday-activities-effortless.

Nissenbaum, H (2010) Privacy in Context: Technology, Policy and the Integrity of Social Life. Stanford, California. Stanford University Press.

O'Brien, D (2020) Data Privacy or Data Protection Day? It's a human right either way. Electronic Frontier Foundation. https://www.eff.org/deeplinks/2020/01/data-privacy-or-data-protection-day-its-human-right-either-way.

OECD.org. (2020, April 23). Tracking and tracing COVID: Protecting privacy and data while using apps and biometrics. https://www.oecd.org/coronavirus/policy-responses/tracking-and-tracing-covid-protecting-privacy-and-datawhile-using-apps-and-biometrics-8f394636/.

Parliamentary Joint Committee on Intelligence and Security, 2019. *Advisory report on the Identity-matching Services Bill 2019 and the Australian Passports Amendment (Identity-matching Services) Bill 2019.* Canberra, Australia: Commonwealth of Australia.

Personal Data Protection Act 2012 – Singapore Statutes Online. (n.d.). Singapore Statutes Online. https://sso.agc.gov.sg/Act/PDPA2012.

Privacy Act 1988. (n.d.). Australian Federal Register of Information. https://www.legislation.gov.au/Details/C2021C00024.

Srimoolanathan, A. (2020, November 19). Impact of Biometrics on Evolving Security Industry in the COVID Era and Beyond. Frost & Sullivan. https://ww2.frost.com/frost-perspectives/impact-of-biometrics-on-evolving-security-industry-in-the-covid-era-and-beyond/.

The Illinois Biometric Information Privacy Act ('BIPA'): When Will Companies Heed the Warning Signs? (2020). The National Law Review. https://www.natlawreview.com/article/illinois-biometric-information-privacy-act-bipa-when-will-companies-heed-warning.

Williams, S. (2020, May 18). What Will Privacy Look Like in a Post-COVID-19 World? CMSWire.Com. https://www.cmswire.com/digital-experience/what-will-privacy-look-like-in-a-post-covid-19-world/.

IV

Prosperity

This final section concludes on how such alternative strategies may lead to a more sustainable future with digital transformation aiding in the process. While the pandemic has created job losses in certain industries, it has created new opportunities in others. Thought leadership in areas such as the future of food, sustainable development goals and achieving the circular economy equips mankind with the ability to achieve prosperity.

CHAPTER 14

A Strategic Architecture for Post-Pandemic Corporate Prosperity

Mark Esposito[1], Alessandro Lanteri[2] and Terence Tse[3]

[1]*Hult International Business School, Cambridge, Massachusetts; Arizona State University, Phoenix, Arizona*
[2]*Hult International Business School, Dubai, UAE; ESCP Business School, Turin, Italy*
[3]*Hult International Business School, London, United Kingdom*

CONTENTS

DOI: 10.1201/9781003148715-14

14.1 INTRODUCTION

The COVID-19 pandemic has been a major disruptor for tried and tested approaches to strategy. More specifically, it acted as an accelerator of two emerging approaches: digital transformation (Soto-Acosta, 2020) and sustainability (Barbier and Burgess, 2020). Even as the pandemic still impacts our lives and livelihoods, companies are stepping up their efforts and new imperatives and business models are emerging every day, as a form of validation of new organisational practices and norms. The pandemic has not only changed our perception of the world but equally accelerated our adoption to new forms of work and production which are becoming visible by the day.

A new survey reveals that 38% managers answered, "we need to re-evaluate" and 31% answered "we are taking steps to change", when responding to how the coronavirus event is affecting their decisions on digital transformation (EY, 2020). In another study, 70% of executives surveyed in Austria, Germany and Switzerland claimed the pandemic had pushed them to accelerate the pace of digital transformation within their firms (Malev, 2020). In a survey of almost 1,600 CEOs from 83 countries, 25% of the respondents strongly agreed that "climate change initiatives will lead to significant new product and service opportunities for [their] organisation" (PWC, 2020). And in a private conversation with a senior McKinsey's partner, it was stated that most of the firm's clients have witnessed an acceleration of roughly 6–7 years on an average, on the efforts of digital transformation across industries, sectors and geographies. So, the change did not just allegedly happen, it did happen from the experience of what we have been able to assess in our practice of educators, consultants and advisors.

These findings are not surprising; companies that are more aggressive in digitising their activities are more likely to end up achieving superior economic performance (Bughin et al., 2017) and those that actively manage a wide range of sustainability indicators are better able to create long-term value for all stakeholders (Funk, 2003).

How do business leaders seize these opportunities?

14.2 A STRATEGIC ARCHITECTURE

While the challenges with new normal are always related to the ability to identify existing frameworks that could help and the synthesis of some of the issues we are facing, in this chapter we present two research-based frameworks of emerging approaches to strategic decision-making that jointly help decision-makers to identify growth opportunities and successfully organise to pursue them in the turbulent, post-pandemic world economy. These frameworks were originally designed by aspiring to future fitness, in integrating adaptable and flexible thought processes in their functioning. They were designed as non-exhaustive instruments, purported to grasp the flow of the times in which we are living, rather than advocating seminal theory within three levels of critique utilised to best capture the entropy of events, in an ever-complex world, where linear models of distribution are less and less functional to the purpose of prediction and forecast.

The macro-level of analysis taps on the contextual need to assess the environment and the nature of the forces that are shaping the conditions for business and public policy to operate. From the scanning of the externalities to the detection of grassroot movements, to the discovery of growth models undetected by traditional economic indicators (Esposito, Tse, Soufani, 2018), this level of analysis is needed to determine the magnitude and direction of the trends.

The meso-level of analysis refers to the institutional and intermediated level of the economy in which a necklace of entities is operating as connectors between forces at the macro and execution on the micro. The meso-level provides opportunities for the formulation of strategy, ideation of business models which probe markets and reconfigure systems. It is an important aspect of how systems become purposeful.

The micro-level involves the most active agents in the economy, such as consumers and individuals, who make decisions on the foundations of limited information, sentiments and access to opportunities. It is where markets get validated and where decision-making blends theoretical assumptions with projectable actionability and where the socioeconomic system enables itself.

Our frameworks (Figure 14.1) combine the three levels of critique in a practical but research-rich rigorous design, aimed at facilitating the transformation of systems by the deeper understanding that emerges from the nature of the trends as well as from the ideation of how the trends can shape strategy. It is within this assumption that we believe that a dual concurrent digital and sustainable transformation can be enacted.

The first of these frameworks, DRIVE (Tse and Esposito, 2017), is a macro-level framework that maps the contextual megatrends shaping

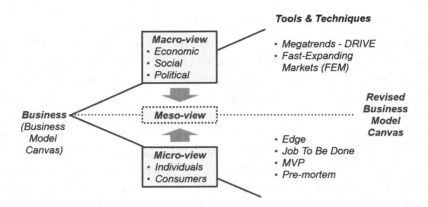

FIGURE 14.1 A way to identify growth opportunities.

the socioeconomic ecosystem and the trajectories emerging from it. The second, CLEVER (Lanteri, 2019), is a meso-micro-level framework that describes the drivers of successful strategy implementation and provides access to a series of actionable levers which can turn vision into reality.

14.3 UNDERSTANDING HOW 'DRIVE' BRINGS THE MACRO-LEVEL OF ANALYSIS TO LIFE

During the research conducted, we have observed, pivoted and researched several points of junction, where the current trends of the present set the trajectory for events to occur in the future, and we believe that these elements carry significance. In the vast ocean of knowledge, information and data, we find that there are five undercurrents that can give us some directions as to how the future would unfold. These five paths aren't exclusive or exhaustive. Most are likely just a minimal representation of all undercurrents that interplay with the shaping forces of our societies, but with this in mind, we have narrowed the research down to five megatrends, which we call the DRIVE framework, consisting of the following:

- Demographic and social changes
- Resource scarcity
- Inequalities
- Volatility, scale and complexity
- Enterprising dynamics

We believe that each of these megatrends is unique in its own right, but, in combination, they can present a fairly comprehensive picture of what the future holds and help us picture a future in the making. Let's look at them under each individual's lens, to best understand the overall purpose of the conceptual framework.

14.3.1 Demographic and Social Changes

At the centre of any future state are people. It is now a known fact that the populations in the developed world are both shrinking and aging, partially a result of the combination of very high life expectancy and very low fertility rates. This is particularly visible if we tap on the visuals of any population pyramid today, which is an easy, albeit serious, effort to envision how populations are distributed by gender and segments of age from 0 to 85+. Interestingly, it was called a pyramid because it implied that children, the base of the population pyramid, are more numerous than the elderly, hence the renowned shape. But from a glance at the global population pyramid today, versus comparison of the population pyramid in 1970, the differences are astounding. Children are yet the most populous group in our societies, but only by marginal units. The pyramid is changing in shape and converging more into a dome, so to say—fatter at the bottom up to 50 years of age and thinner from 50 and above but definitely far from resembling a pyramid at best. Projections show that the majority of populations of developed countries will be over 40 years old by 2030, with Japan reaching an average of 52 by then, followed by Italy and Germany. Less obvious though, the same is happening in the developing economies. Take China, for instance. Even though the Middle Kingdom has a younger population (35.4) today than the United States (37.4), by 2030 this will be 42.1 compared to 39.5—China's aging population is going to "catch up" with the developed world (Roland Berger Strategy Consultants, 2015). Most emerging economies remain young, however.

The most youthful are those of the sub-Saharan region. For example, the average age of Nigerians is 15 years old in 2013. Yet, by 2030, the average age of the country would only be 15.2 years old!

The implications of such demographic shifts can be huge and destabilising at best. While the developed world will be facing challenges such as maintaining and sustaining the social security systems and financing pensions, because of an influx of its elderly population, the developing ones will have to be able to provide education and access to basic health care to youngsters as well as create jobs for them. On a global basis, as people are getting older, the world's population will plateau for the first time in human history. A worrying implication is that there will be insufficient labour power to support the old-age pensioners (Dobbs, Ramaswamy, Stephenson, & Viguerie, 2014). With the labour pool shrinking, the only way to maintain economic growth is to continuously invest in raising productivity and competitiveness, away from natural resources and more toward models of economic efficiency that do not resort to intense factors of production, to extract value. This not only requires companies to increase their efforts – and capital – in making investments, at least in the medium term, but this could also put governments in a dilemma. What is needed are more liberal economic and business policies, as past experience has shown that productivity significantly increases in industries that are unprotected and can freely compete, coupled with a modern fiscal system that reinvests taxes into logistics, mobility, and public services. Yet in times of uncertainties, governments are more inclined to erect trade barriers and pursue protectionism, or to alienate the electorate with short-term austerity measures, which distance countries even further from this template of increased productivity – the exact opposite to what should be done to make the future better (Godin & Mariathasan, 2014).

As people are getting older, they are also converging: more and more people are living in cities. In 1950, the urbanisation ratio in China was 13%. Today, roughly half of the country's population lives in cities and the government plans to push that to 70% by 2025 (Johnson, 2013). There is a good reason for having more city dwellers: greater urbanisation is shown to have a positive effect on gross domestic product (GDP) development. In fact, countries with the highest urbanisation ratios are showing the greatest GDP per head of the population (Johnson, 2013). Moreover, countries with high populations and levels of urbanisation tend to have the strongest GDP growth. And while this may seem to be challenged by the progressive hybridised work agenda around the world, there is reason to believe that the recent developments due to COVID-19 will not alter or change the course of action of an inexorable intensification of urbanisation in the years to come.

14.3.2 Resource Scarcity

Urbanisation and the continuous growth of world's population size would put a lot of pressure on the use of resources because cities and resources to run them are always co-related. But after years of resource exploitation, dating back from colonialism to our days, resources are now rapidly depleting and dropping in quality, or at least the availability cycle is undermined (Tse, Esposito, & Soufani, 2014). Resource scarcity, inevitably, represents another megatrend. Many people relate resource scarcity to energy, thinking that the world would run out of fossil fuels soon. Fossil fuels are indeed nonrenewable.

But there are reasons why the worry is unwarranted. First, there is a huge amount of reserves that are yet to be inventoried. Second, there are always wind and solar energy, which, in addition to the associated sensitivity to shifting to clean energies, have become largely available also because of the climatic changes. Water and food, on the other hand, are different stories. At the current consumption rate, according to the World Wildlife Fund (n.d.), it is possible that by 2025 two-thirds of the world's population may face water shortages. It is not just a matter of not quenching thirst.

As cities becoming ever more urbanised, the increased requirements on sanitation would put further pressure on the use of water and beyond, in what we see as strings of co-related shortages. At the same time, most companies' value chains would also be deeply affected by water scarcity, across regions, but more so, where manufacturing is still the key driver of production. It takes, for instance, an average car containing about 2150 pounds of steel, this would mean over 300,000 L of water is needed to produce the finished steel for just one car (Grace Communications Foundation, n.d.).

Lately, circular economy, also known as the 'cradle-to-cradle' model, is gradually gaining more and more traction as the model of the future, and, hopefully, its inherent regenerative modus operandi will gain more and more ground, carving space for a needed shift to circularity. This is where we believe there will be a dedicated focus towards a sustainability transformation, which will rise from the ashes of the phoenix, in this case, our old models of productions that have been grounded to zero during the pandemic.

14.3.3 Inequalities

The world seems to have woken up to the issue of inequality when the French economist Thomas Piketty (2013) argued in his bestseller that the unequal distribution of wealth in the developed countries has become more so in recent years and since then, it didn't get any better. The increasing pressure on the systems and the gap becoming inexorably larger, coupled with structural dysfunctions caused by the degrees of implications inferred by the pandemic, our world is more unevenly distributed now than ever. In the past 30 years, the incomes of the wealthiest have surged into the stratosphere (and the higher up in the income hierarchy one is, the greater the increase has been), while the incomes of the large majority have stagnated. This has led to a level of inequality in wealth in the developed world not seen since the eve of the Great Depression (Piketty, 2013). Income inequality – while a serious dysfunction of how wealth should be distributed – is, unfortunately, not necessarily the most acute of the socioeconomic gaps we have analyzed. In fact, just as worrying is that the phenomenon that the middle-income class is gradually disappearing. In the United States, while productivity and GDP have continued to grow, middle-income earners have been making less over time: the percentage of households earning 50% of the average income has decreased from 56.5% in 1979 to 45.1% in 2012 (Bernstein & Raman, 2015).

Worst yet, this 'hollowing out' is not only confined to the United States; the same phenomenon is also observed in 16 European countries around the same period (Goos, Manning, & Salomons, 2014). This, as it turns out, is the consequence of automation and computerisation (more below). While the middle class disappearing may suggest a rebalance of wealth toward either poor or rich people, the story behind may present bleaker aspects and nuances. Middle class in modern civilisations is the bearing engine of our economic outputs. In dry terms, it is where GDP is ultimately produced. If the middle class slims itself to the level of becoming marginally relevant, repercussions in the standards of living must be anticipated as likely to happen. The framework doesn't exclusively mention income inequality as the sole form of inequality but we tend to believe that a series of socio-economic inequalities are on the rise, even if what seems alleged to be non-financial matters, such as life expectancy, access to education, gender opportunities, or simply access to healthcare. This is why we believe that the repercussions of widening inequalities in various dimensions of our societies, especially when very little effort has been made to stop it, could greatly impact our future, inferring a trajectory we hope to be able to correct, before its full deployment in our societies.

Indeed, the current pandemic is most likely going to drive inequality further as a result of the so-called K-shaped recovery: large companies and public-sector institutions with direct access to government and central bank stimulus packages will make some areas of the economy recover fast but leave others out. This comes at the expanse of small and medium-sized enterprises, blue-collar workers, and the dwindling middle class (Bheemiah, Esposito and Tse, 2020). It must be noted that technologies alone are not necessarily the solution to reverse this situation. Technologies needed to be complemented with the right government policies and company practices to address these inequality concerns.

14.3.4 Volatility, Scale and Complexity

The world has not been a tremendously exciting place in terms of economic growth for most of the past two millennia. Such growth started to accelerate only after the Industrial Revolution. The invention of steam power in the mid-eighteenth century, followed by the emergence of an internal combustion engines, electricity, and household plumbing about a century later, brought significant economic growth to the world's economy (Gordon, 2012).

More recently, the revolution that caused our economy and productivity to take off once again was the advent of Internet technologies (Brynjolfsson & McAfee, 2014). One of the most observed writers in this field, Jeremy Rifkin (2011), speaks about a third industrial revolution when the fledging of existing platforms or engines – namely, the energy, communication, and mobility drivers – converge. Indeed, what makes online technology different from – and far more powerful than – other revolutionary technologies within this integrated view of converging models in the past is the fact that it acts as a glue to different types of devices and technologies, often leading to something novel. Things have changed since then and from the incremental nature of technologies, hassled by the concerns raised by Moore's Law, we see today that new forms of deeptech are generating more and more combinatorial outputs and these outputs are giving rise to new priorities ahead.

One outcome emerging from the combined use of these technologies is robotics, which has recently reached a point of intelligence at which they will be able to help humans in every kind of industry, in ways previously unimaginable. The penetration of robotics into regular activities is increasing as we speak, with examples of robotics applied also to biology, in what futurists call the 'fourth industrial revolution,' where the combination of biology, robotics, and digital may become a new integrated reality of products and services.

The technological tipping point has been the development of advanced sensors. This has given robots the ability to sense and interpret the world around them (Williams, 2016). Just as important is the arrival of cloud computing. Previously, robots would have to learn all by themselves as individual entities. With the possibility of linking them all up to a single source, learning by one can be easily shared with the others. This puts robotics on a very steep learning curve and hence faster developments (Ross, 2016). The rise of robotics could, like the earlier Internet revolution, lead to the development of new ecology around robotics, providing plenty of new opportunities. The fast-paced technology development would require businesses to be adaptive but above all, capable of synthesising and integrating elements that may learn quicker than the programmer would have ever imagined. And we are just getting started.

14.3.5 Enterprising Dynamics

Technologies are not just fundamentally changing the way we live; they are also reshaping how businesses work. While our research on volatility demonstrates a clear orientation towards combinatorial modus operandi, other parts of the world – namely the emerging economies – are living this reality by contextualising many of these technologies to serve local markets. To many people in the West, China is often viewed as a country unable to innovate and prone to acquiring existing technology from others. This reflects perhaps a natural response that China's stellar economic rise is partly due to the country's inclination to assimilate existing Western products and, worse yet, sell fake goods. Yet such a view underscores the fact that many people confuse innovation with invention. The Middle Kingdom may not be very good at coming up with products based on its ideas, but this does not mask the truth that in certain areas it has been doing business with such a level of novelty that Western companies should take as reference. Whereas businesses in the West and Japan remain strong in engineering- and science-based innovation, Chinese firms excel in the customer-focused and efficiency-driven sorts (Woetzel et al., 2015). This might sound trivial at first glance, but we imagine that it would be rather difficult in the United States or the United Kingdom for online platforms to extend into financial services without running into regulatory concerns or competition from incumbents such as banks (Tse, 2015). Naturally, Western companies do not standstill. Companies of all sizes are taking on new and different entrepreneurial dynamics. For instance, a London-based translation company is now much more of an information technology (IT) integrator with a small set of translation activities. This is not only because computers and software can now do most of the translation; this is also due to large multinational clients demanding full IT integration to their systems. In this instance, these corporations require the translation company to translate all of their invoices into different languages automatically, and all of these must be conducted seamlessly on their platforms. In some cases, while the business model is new, the customers can often tweak it to add a new dimension to the business model, customising it in some way at every interaction. Compared to the past, one would know if there is an actual demand, only when it gets to the stage of getting ready to sell the product to customers. But the entrepreneurs would have to come up with a concept, build a prototype, raise the funding, conduct market research, manufacture and hold inventory, and find sales channels and negotiate with them, all of which consume an enormous amount of time and money. As more and more businesses are now conducted online, it can be expected that there will only be more dramatic and speedier changes in business models and value chains in the years to come.

14.4 DISCOVERING THE CLEVER WAY TO ENACT ON THE MESO-/MICRO-LEVEL OF THE ECONOMY

Moving the level of analysis to the meso-level, CLEVER is a framework of strategic drivers that is the forces that offer strategic advantages and – therefore – trigger strategic change. These six drivers are:

- Collaborative intelligence
- Learning systems
- Exponential technologies
- Value facilitation
- Ethical championship
- Responsive decision making.

The first three elements of the framework represent the 'hard' features – the technological components – the following three capture decision-making and managerial styles. Table 14.1 reports the definition and strategic advantages of each driver, as well as illustrative examples and counterexamples.

14.4.1 Collaborative Intelligence

Collaborative Intelligence refers to the intelligent coordination of a broad range of skills. It is a strategic driver that allows outcomes and performances that are not simply better, but otherwise altogether impossible.

TABLE 14.1 Definition and Strategic Advantages of Each Driver in the CLEVER Framework

	Collaborative Intelligence	Learning Systems	Exponential Technologies	Value Facilitation	Ethical Championship	Responsive Decision Making
Definition and strategic impact	The intelligent coordination of increasingly broad ranges of skills. It allows outcomes and performances that are not simply better, but otherwise altogether impossible.	The use of data analytics empowers decisions that are not only faster and more accurate, but also continuously become ever faster and more accurate.	Leveraging technologies that improve at an accelerating rate (in terms of cost and performance) and support rapidly increasing productivity	The coordination of transactions among users. It allows faster growth, with less capital and fewer employees, and so achieving higher profitability.	The pursuit of an aspirational reason for being (beyond money). It helps seize new opportunities, give meaning to peoples' lives, attract talent, successfully innovate, navigate change, minimise risks, reduce costs, increase revenues, and earn a societal license to operate.	Approaching strategic decisions as discovery and learning processes of fast-changing competitive conditions, which helps organisations to respond in a timely and progressively more effective fashion.

(Continued)

TABLE 14.1 (Continued) Definition and Strategic Advantages of Each Driver in the CLEVER Framework

	Collaborative Intelligence	Learning Systems	Exponential Technologies	Value Facilitation	Ethical Championship	Responsive Decision Making
Illustrative examples	Wikipedia's use of the crowd; Amazon's robot-human warehouse management; chess players in freestyle competitions; financial roboadvisors; robosurgery.	Google's search predictions; email spam filtering; Alibaba's Taobao; Ant Financial's data-driven new insurance products; IBM's Watson; Netflix's recommendation system; Dubai Airports' passengers' flow; Harley Davidson's web marketing.	Uber's use of mobile and GPS; Jio's transition from oil to mobile; Instagram's use of personal cameras; Google Glasses 2.0; ubiquitous computing; additive manufacturing and 3-D printing; from the Human Genome Project to consumer DNA test kits.	Email; Airbnb; Baidu; bitcoin/ blockchain; John Deere's MyJohnDeere; the World Economic Forum at Davos.	Unilever's Sustainable Living Brands; social enterprises and B-Corp; UN's Sustainable Development Goals; Grameen Bank; Business Roundtable's Statement on the Purpose of a Corporation; Patagonia's discouraging customers from buying new garments; Environmental, Social, and Governance (ESG) investments, Socially Responsible Investments (SRI), and impact investing.	New Balance's always in beta; Rent The Runway's launch; Dropbox's market test; PayPal's pivot from Palm Pilots to email; Amazon's always day one; the Agile Manifesto; lean startup methodologies; design thinking.
Counterexamples	Tesla's excessive automation; retail apocalypse.	-	AT&T's sale of its mobile operations in the 80's and subsequent repurchase; Kodak's failure to exploit digital photography; Google Glasses 1.0.	Nokia's failure to develop a platform open to third parties.	BP's Deep Horizon catastrophe; Facebook's Cambridge Analytica scandal; Uber's former CEO scandals.	Segway's initial dud; failure of 80% of new product launches.

A good example of the power of Collaborative intelligence is the online open encyclopaedia Wikipedia. Its content originates from a vast community of contributors from different backgrounds, each of whom offers their knowledge on a broad range of topics. Some contributors volunteer their editing and reviewing skills. Wikipedia also assigns separate tasks to automated bots and humans, so that each performs the tasks they are most qualified for. For example, bots verify whether new content is plagiarised. They can rapidly compare the percentage of text that identical to that already published elsewhere, and quickly raise a warning if the percentage is too high. Such important work would be too time-consuming for humans. However, bots are incapable of editing the text to avoid plagiarism. So, if necessary, a human can step in and execute an informed correction. The diverse skillsets of many humans (a crowd) and bots achieved what human experts alone could not.

Collaborative intelligence does not emerge by simply bundling together different skillsets but ultimately depends on the process of coordinating them (Malone, 2018; McAfee, 2010). Some examples of Collaborative intelligence are the combination of the skills of humans and machines (Daugherty and Wilson, 2018), the power of the crowds with diverse skillsets, experiences, expertise, and origins (Surowiecki, 2005), and open innovation (Chesbrough, 2003) that leverages knowledge and resources that are both internal and external (e.g., consumer feedback, competitors, universities) to an organisation.

14.4.2 Learning Systems

Learning Systems consist of the use of data analytics to process large amounts of data to gain insight and make superior automated decisions. Even with the simple forms of learning and automation (Davenport and Ronanki, 2018; Iansiti and Lakhani, 2020) in use nowadays, it is a strategic driver that empowers decisions that are not only faster and more accurate, but also continuously become ever faster and more accurate.

At the core of this strategic driver are machine learning (ML) algorithms that explore patterns among vast stores of diverse data – ranging from demographic information to spending habits, from social interactions to online search history, from geolocation to personality... – and identify statistical correlations between inputs and some relevant outputs. This way, when they are given new inputs they can quickly and cheaply predict the corresponding outputs (Agrawal *et al.*, 2018).

A good illustration of Learning Systems is the video content producer and distributor Netflix. They gather large amounts of data about the viewing behaviour and preferences of their subscribers. Netflix's algorithms can then predict what content each viewer is likely to enjoy each time they login. These algorithms are so accurate that 80% of the content streamed on the platform comes from recommendations (Plummer, 2017).

Other content producers shoot a pilot episode before producing new series, in order to test it with its target audience, and gather feedback that helps predict future viewership. Netflix does not shoot pilots. It uses data to choose what new content to produce. Viewers' ratings of different content, genres, and actors suggested the very successful remake of the 'House of Cards' series (Carr, 2013). The success rates of traditional pilot-tested TV shows are 30%–40%. Netflix original shows' are 80%(Orcan Intelligence, 2018).

14.4.3 Exponential Technologies

Exponential Technologies consists of leveraging of technologies whose performance improves at an accelerating rate and so support rapidly increasing productivity. Digital technologies—such as digital sensors, computing power and blockchains—improve very rapidly and the performance improvement of each new version of the technology is greater than all the cumulated improvements until that point.

For example, in November 2019, the best camera phones on the market had a resolution ranging between 12 megapixels (MP) and 48 MP (Wired 2019). Soon thereafter a 108 MP camera phone was released (Kelion, 2019). Similarly, 5G connectivity is expected to transmit data one hundred times faster than the previous 4G standard. As 5G is being rolled out, 6G is already poised for release, with an anticipated connectivity one hundred times faster than 5G.

Besides the speed of release and the accelerated improvements, digital technologies integrate into and reinforce each other, a process called convergence that results in rapidly increasing productivity (Camerona et al., 2005). These accelerating digital technologies are poised to transform "the way we work and consume (commerce), our well-being (health), our intellectual evolution (learning), and the natural world around us (environment)" (Segars, 2018, p. 4). If digital technologies continue improving at this accelerating rate, over the next twenty years we will experience a degree of innovation comparable to that experienced over the last century (Kurzweil, 2001).

14.4.4 Value Facilitation

Value Facilitation refers to the coordination of transactions among users, via a platform (Parker and Van Alstyne, 2016). In market exchange, value is created between buyers and sellers completing transactions. These platforms do not create value; they facilitate transactions that create value. Value Facilitation is a strategic driver that allows faster growth, with less capital and fewer employees, and so achieving higher profitability. This way, value creation is largely curated by users, often using spare capacity or underused assets. Since users create value for each other (Katz and Shapiro, 1986), the more users participate in a platform, the more valuable the platform becomes. Seven of the world's twelve largest companies by market capitalisation operate a business model of this type (Hirt, 2018).

For example, Airbnb does not directly engage in value-creating exchanges. It facilitates exchanges that create value between hosts and guests. Launched in 2008, Airbnb listed 1 million properties by 2015 (Hagiu and Rothman, 2016), and by 2020 it surpassed seven million listings (Airbnb, n.d.). Marriott, the largest hotel company in the world, took fifty-eight years to get to one million rooms and reached 1.2 million rooms only in 2020 (Trejos, 2018).

14.4.5 Ethical Championship

Ethical Championship refers to the pursuit of an aspirational reason for being beyond money. It is a strategic driver that helps seize new opportunities, give meaning to peoples' lives, attract talent, successfully innovate, navigate change, minimise risks, reduce costs, increase revenues, and earn a societal license to operate. All this is reflected in the financial bottom line, as companies with a purpose outperformed the S&P 500 by a factor of 14 between 1998 and 2013 (Sisodia, Sheth, & Wolfe, 2014).

Most people believe that "capitalism, as it exists today, does more harm than good in the world" (Edelman, 2020, p. 12). Curbing the negative side effects of capitalism by embracing Corporate Social Responsibility, or CSR (Carroll, 1991) or meeting a triple bottom line (Elkington, 1998) is not enough. The next, bolder step, requires the deliberate pursuit of dual goals—financial sustainability and positive social and environmental impact (Battilana et al., 2019; Dacin et al., 2011)—to ensure business becomes a force for good.

For example, consumer goods multinational Unilever has a portfolio of over 400 brands, including household names like Axe, Knorr, Lipton, Magnum. In the early 2000's it began transforming some of its brands, associating them with a purpose. The first such case is Dove, a skincare brand, which started celebrating the natural beauty of every woman, instead of promoting idealised beauty standards embodied by airbrushed models, which cause many girls and women to feel insecure about their appearance. Dove stopped using professional models for its campaigns and instead hired women in a wide range of skin colours and body shapes, to pass the message that real women, with stretchmarks and all, are beautiful. Six months after launch, sales went up 700%. Today, Dove is Unilever's biggest brand. Unilever kept investing in brands with purpose, which it calls 'sustainable living brands.' By 2017, more than half of Unilever's top-performing brands were sustainable living brands. Moreover, these brands with purpose grew 46% faster than the other brands and account for 70% of Unilever's revenue growth.

14.4.6 Responsive Decision Making

Responsive decision making consists of approaching strategic decisions as discovery and learning processes in fast-changing competitive conditions. It is a strategic driver that helps organisations to respond in a timely and progressively more effective fashion to uncertainty. Some examples of responsive decision making are dynamic capabilities (Teece *et al.*, 1997), a learning culture (Garvin, 1993), the lean startup (Blank, 2013; Ries, 2011), hypothesis-driven learning (Eisenmann *et al.*, 2014) and design thinking (Brown, 2008; Martin, 2009).

To remain successful in fast-changing competitive landscapes, assets must be reconfigured and redeployed and both internal and external structures must be redesigned. Yet, it is not enough to reorganise. Reorganisation itself must be reimagined to fit turbulent environments.

For example, GE developed the Durathon battery following an agile approach (Blank, 2013). It began approaching potential customers to explore their needs and expectations, iteratively developing and testing prototypes that, while not ready for the market, helped gather invaluable feedback. In the process, GE abandoned some of its original target customers, identified new market niches, and validated the viability of its Durathon battery before its manufacturing plant was ready. If it had followed the traditional waterfall approach, it would have built a production facility to roll out the new product and then it would have started selling it, only to find out that some of its target customers were 'wrong,' and so it would have wasted large resources and then had to fix its offering, at a much greater cost and possibly too late.

14.5 CONCLUSION

To successfully navigate the upheaval caused by the COVID-19 pandemic, business leaders must reconsider their strategy. Two of the key strategic questions they must answer are: where their company will play and how will it win. This chapter provides two research-based frameworks to answer these two critical questions.

When answering the first of these questions, DRIVE assists decision makers in tracking the megatrends that shape the competitive landscape and in identifying major emerging opportunities, which traditional tools for strategic planning might only capture when it is too late. For the second question, CLEVER maps the drivers of strategic advantage decision makers should leverage to maintain a competitive edge. Jointly, the two frameworks constitute the foundation of a new strategic architecture that empowers firms to identify growth opportunities and successfully organise to pursue them in the turbulent, post-pandemic world economy.

Although they've only been published for less than 5 years, our frameworks have now been taught in several graduate programs in business schools around the globe and have become the cornerstone of many executive education programs and consulting projects. Despite these accolades, these frameworks do not aspire to be perfect. In turbulent times, with multi-causal patterns and nonlinear trajectories of change, the future cannot be predicted. So, even equipped with the DRIVE and the CLEVER frameworks, decision makers might still face sudden black swans and wildcards that lead to unexpected circumstances and rapidly shifting competitive scenarios.

Frameworks should instead be judged for their usefulness. In the years spent researching and refining our frameworks, we have witnessed firsthand how they have been applied in dozens of companies across industries and geographies. Business leaders can use the two frameworks to understand where strategic opportunities emerge, to find inspiration for strategic action, and so ensure their organisations become stay future-ready in the post-pandemic era.

REFERENCES

Agrawal, A., Gans, J., & Goldfarb, A. (2018). *Prediction machines: The simple economics of artificial intelligence*. Cambridge, MA: Harvard Business Review Press.

Airbnb (n.d.). *About us*. Retrieved April 20, 2020. https://news.airbnb.com/about-us

Barbier, E. B., & Burgess, J. C. (2020). Sustainability and development after COVID-19. *World Development, 135*, 105082.

Battilana, J., Pache, A.C., Sengul, M., Kimsey, M. (2019). The dual-purpose playbook. *Harvard Business Review, 97*(4), 124–133.

Bernstein, A., & Raman, A. (2015, June). The great decoupling: An interview with Erik Brynjolfsson and Andrew McAfee. *Harvard Business Review*, 2015, 66–74. https://hbr.org/2015/06/ the-great-decoupling

Bheemiah, K., Esposito M., & Tse, T. (2020). K-Shape Recovery, World Economic Forum Blog, retrieved February 2021.

Blank, S. (2013). Why the lean startup changes everything. *Harvard Business Review, 91*(5), 63–72.

Brown, T. (2008). Design thinking. *Harvard Business Review, 86*(6), 84–92.

Brynjolfsson, E., & McAfee, A. (2014). *The second machine age: Work, progress, and prosperity in a time of brilliant technologies.* New York, NY: W. W. Norton.

Bughin, J., LaBerge, L., & Mellbye, A. (2017). *The case for digital reinvention.* https://www.mckinsey.com/business-functions/mckinsey-digital/our-insights/the-casefor-digital-reinvention

Camerona, G., Proudman, J., & Redding, S. (2005). Technological convergence, R&D, trade and productivity growth. *European Economic Review, 49*, 775–807.

Carr, D. (2013, February 24). Giving viewers what they want. *The New York Times.* https://www.nytimes.com/2013/02/25/business/media/for-house-of-cards-using-big-data-to-guarantee-its-popularity.html

Carroll, A. (1991). The pyramid of corporate social responsibility: Toward the moral management of organizational stakeholders. *Business Horizons, 34*(4), 39–48.

Chesbrough, H. W. (2003). *Open innovation: The new imperative for creating and profiting from technology.* Cambridge, MA: Harvard Business School Press.

Dacin, T., Dacin, P., & Tracey, P. (2011). Social entrepreneurship: A critique and future directions. *Organization Science, 22*(5), 1203–1213.

Daugherty, P. & Wilson, J. (2018). *Human + machine.* Cambridge: Harvard Business Review Press.

Davenport, T.H. & Ronanki, R. (2018). Artificial intelligence for the real world. *Harvard Business Review, 96*(1), 108–116.

Dobbs, R., Ramaswamy, S., Stephenson, E., & Viguerie, S. P. (2014, September). Management intuition for the next 50 years. *McKinsey Quarterly.* http://www.mckinsey.com/business-functions/strategy-and-corporate-finance/our-insights/management-intuition-for- the-next-50-years

Edelman (2020). *Trust Barometer 2020.* https://www.edelman.com/trustbarometer.

Eisenmann, T., Ries, E., & Dillard, S. (2014). *Entrepreneurship reading: Experimenting in the entrepreneurial venture.* Cambridge, MA: Harvard Business Publishing.

Elkington, J. (1998). *Cannibals with forks. The triple bottom line of 21st Century.* Oxford: Capstone.

Esposito, M., & Tse, T. (201). DRIVE: The five megatrends that underpin the future business, social, and economic landscapes. *Thunderbird International Business Review, 60*(1), 121–129.

Esposito, M., Tse, T., & Soufiani, K. (2016). The seminal concept of Fast Expanding Markets. *Thunderbird International Business Review, 59*(1), 5–7.

Esposito, M., Tse, T., & Soufiani, K. (2018). The circular economy: An opportunity for renewal, growth, and stability. *Thunderbird International Business Review, 60*(5), 725–728.

EY (2020). *How Do you Find Clarity in the Midst of a Crisis?* https://www.ey.com/en_us/ccb/how-do-you-find-clarity-in-the-midst-of-covid-19-crisis

Funk, K. (2003). Sustainability and performance. *MIT Sloan management review, 44*(2), 65.

Garvin, D., (1993). Building a learning organization. *Harvard Business Review, 73*(4), 78–91.

Godin, I., & Mariathasan, M. (2014). *The butterfly defect: How globalization creates systemic risks, and what to do about it.* Princeton, NJ: Princeton University Press.

Goos, M., Manning, A., & Salomons, A. (2014). Explaining job polarization: Routine-biased technological change and offshoring. *American Economic Review*, 104, 2509–2526.

Gordon, R. (2012, September). Is US economic growth over? Faltering innovation confronts the six headwinds. *Centre for Economic Policy Research Policy Insight*, 63, 1–13. Retrieved from http://www.cepr.org/ sites/default/files/policy_insights/PolicyInsight63.pdf

Grace Communications Foundation. (n.d.). The hidden water in everyday products. Retrieved from http://www.gracelinks.org/285/the- hidden-water-in-everyday-products

Hagiu, A., & Rothman, S. (2016). Network effects aren't enough. *Harvard Business Review, 91*(3), 102–108.

Hirt, M. (2018). If you're not building an ecosystem, chances are your competitors are. *McKinsey & Company.* https://www.mckinsey.com/business-functions/strategy-and-corporate-finance/our-insights/the-strategy-and-corporate-finance-blog/if-youre-not-building-an-ecosystem-chances-are-your-competitors-are

Iansiti, M., & Lakhani, K, R. (2020). Competing in the age of AI. *Harvard Business Review, 98*(1), 60–67.

Johnson, I. (2013, June 15). China's great uprooting: Moving 250 million into cities. *New York Times.* http://www.nytimes.com/2013/06/16/world/asia/chinas-great-uprooting-moving-250-million-into-cities.html?pagewanted=all &_r=0

Katz, M.L., & Shapiro, C. (1986). Technology adoption in the presence of network externalities. *The Journal of Political Economy, 94*(4), 822–841.

Kelion, L. (2019, November 5). *BBC.* Xiaomi smartphone has 108 megapixel camera. https://www.bbc.com/news/technology-50301665

Kurzweil, R. (2001). *The law of accelerating returns.* Kurzweil Accelerating Intelligence. https://www.kurzweilai.net/the-law-of-accelerating-returns

Lanteri, A. (2019). *CLEVER: The six strategic drivers for the Fourth Industrial Revolution.* Lioncrest.

Malev, M. (2020). *Is the coronaviruus pandemic an engine for digital transformation?* https://dmexco.com/stories/is-the-coronavirus-pandemic-an-engine-for-the-digitaltransformation/

Malone, T. (2018). *Superminds: The surprising power of people and computers thinking together*. New York: Little, Brown and Company.

Martin, R. (2009). *The design of business: Why design thinking is the next competitive advantage*. Cambridge, MA: Harvard Business Press.

McAfee, A. (2010). Did Gary Kasparov stumble into a new business process model? *Harvard Business Review*. https://hbr.org/2010/02/like-a-lot-of-people

Orcan Intelligence (2018, January 12). *How Netflix uses big data*. Medium. https://medium.com/swlh/how-netflix-uses-big-data-20b5419c1edf

Parker, G., & Van Alstyne, M. (2016). Platform strategy, in: M. Augier, D.J. Teece (Eds.), *The Palgrave Encyclopedia of Strategic Management*. London: Palgrave Macmillan.

Piketty, T. (2013). *Le capital au XXIème siècle [Capital in the twenty- first century]*. Paris, France: Seuil.

Plummer, L. (2017, August 22). This is how Netflix's top-secret recommendation system works. *Wired*. https://www.wired.co.uk/article/how-do-netflixs-algorithms-work-machine-learning-helps-to-predict-what-viewers-will-like

PWC (2020). 23rd Annual Global CEO Survey. https://www.pwc.com/gx/en/ceo-agenda/ceosurvey/2020.html

Ries, E. (2011). *The lean startup*. Portfolio Penguin.

Rifkin, J. (2011). *The third industrial revolution: How lateral power is transforming energy, the economy, and the world*. New York, NY: St. Martin's Press.

Roland Berger Strategy Consultants. (2015). Roland Berger trend compendium 2030. https://www.rolandberger.com/gallery/trend-compendium/tc2030/content/assets/trendcompendium2030.pdf

Ross, A. (2016). *The industries of the future*. New York, NY: Simon & Schuster.

Segars, A.H. (2018). Seven Technologies Remaking the World. *MIT Sloan Management Review*. https://sloanreview.mit.edu/projects/seven-technologies-remaking-the-world

Sisodia, R., Sheth, J., & Wolfe, D. (2014). *Firms of endearment: How world class companies profit from passion and purpose* (2nd ed.). New Jersey: Pearson FT Press.

Soto-Acosta, P. (2020). COVID-19 pandemic: Shifting digital transformation to a high-speed gear. *Information Systems Management, 37:4,* 260–266.

Surowiecki, J. (2005). *The wisdom of crowds*. New York: Anchor Books.

Teece, D., Pisano, G., Shuen, A. (1997). Dynamic capabilities and strategic management. *Strategic Management Journal, 18*(7), 509–533.

Trejos, N. (2018). The brands and hotel rooms of Marriott International, by the numbers. *USA Today*. https://www.usatoday.com/story/travel/roadwarriorvoices/2018/01/22/brands-and-hotel-rooms-marriott-international-numbers/1053593001

Tse, E. (2015). *China's disruptors: How Alibaba, Xiaomi, Tencent, and other companies are changing the rules of business*. New York, NY: Penguin.

Tse, T., & Esposito, M., (2017). *Understanding how the future unfolds: Using DRIVE to harness the power of today's megatrends.* Lioncrest.

Tse, T., Esposito, M., & Goh, D. (2019). *The AI republic: Building the nexus between humans and intelligent automation.* Lioncrest.

Tse, T., Esposito, M., & Soufani, K. (2014, November 4). How companies can benefit from the circular economy [Web log post]. *California Management Review.* Retrieved from http://cmr.berkeley.edu/blog/2015/11/circular_economy/ World Wildlife Fund (n.d.)

Williams, A. (2016, February 18). *The robotics revolution: Why investment in robots is booming.* City A.M. Retrieved from http://www.cityam.com/234883/the-robotics-revolution-why-investment-in-robots- is-booming

Woetzel, J., Chen, Y., Manyika, J., Roth, E., Seong, J., et al. (2015, October). *The China effect on global innovation.* London, UK: McKinsey Global Institute.

CHAPTER 15

Digital Servitisation and Reverse Logistics Towards a New Circular Economy

Adrian T. H. Kuah[1] and Chang H. Kim[2]

[1]James Cook University Singapore:
The Cairns Institute, Cairns, Australia
[2]The Cairns Institute, Cairns, Australia

CONTENTS

DOI: 10.1201/9781003148715-15

15.1 INTRODUCTION

COVID-19 has changed the way we live and work. While governments have implemented differing extents of lockdown measures to curb regional and national transmission, people are also entering more work-from-home modes and some extents of self-isolation. While the pandemic has contributed to the growth of online businesses and deliveries, it has brought concerns about waste generation. The pandemic triggered considerations how consumption and production patterns can be shifted to a more sustainable way. As a sustainable consumption and production concept was introduced as one of United Nation's Sustainable Development Goals (UN SDGs), businesses have mapped out various strategies on resource extraction, distribution, waste disposal, reuse of products and services to further this quest. Reverse logistics, as a way to realise this quest, has been constantly referred to in literature. However, such a system has not attracted much attention from businesses due to the high cost and complexity of the process. Nonetheless, today, it is noteworthy that more companies have begun to combine digital technologies with their reverse logistics systems to create a new competitive advantage. When a service incorporating digital technologies is applied to a reverse logistics system, it becomes a new solution for businesses to achieve sustainable consumption and production. This chapter set out to shine new light on these emerging areas through an examination of reverse logistics and digital servitisation. Collectively, it will provide useful strategies for businesses seeking to achieve sustainable consumption and production in the post-COVID world by looking at how digital servitisation can be combined within a reverse logistics system.

15.2 COVID-19: A NEW CONTRIBUTOR TO WASTE GENERATION

The COVID-19 pandemic has altered how people live, work and engage with others. One aspect in which COVID-19 differs from the past crises is the fundamental change of behaviour in consumers. COVID-19 is driving changes to consumer lifestyles, where an inherent fear of human-to-human contact has increased, which gradually led to a shift to contactless consumption. Besides, with governments' lockdown measures to tackle the health crisis, people are entering into frequent self-isolation and work-from-home modes. These circumstances are fueling the growth of the "stay-at-home economy" (World Economic Forum, 2020). The rapid rise of the stay-at-home economy market will bring about the rapid growth of contactless services such as e-commerce, video conferencing, online education, and cloud services.

People have changed their consumption patterns as they are transited to increased online entertainment, shopping, and food delivery. Accordingly, online shopping, including the purchase of groceries, has become more prevalent with the onset of lockdown measures in many countries (Donthu & Gustafsson, 2020). The growth of online shopping due to the pandemic has not only minimised the movement of people, and allowed consumers to enjoy the benefits of time, safety, and convenience.

However, such convenience and benefits are turning into the planet's worst nightmare. The stay-at-home economy has significantly increased demand for home delivery services which led to a surge in the generation of unnecessary packaging wastes, including plastics and cardboard boxes (Sharma et al., 2020). The global plastic packaging market size during the pandemic should grow from USD 909.2 billion in 2019 to 1012.6 billion by 2021, at a compound annual growth rate of 5.5%, mainly due to the e-commerce sales (Business Wire, 2020; Markets and Markets, 2020).

The problem is that most of the packaging wastes are discarded after a single use; they are designed to protect the product but are disposed of for the convenience of consumers. If so, it seems obvious that the increase in single-use products like plastics or cardboard boxes will have long-term impacts on the environment. It is worth noting that the widespread practice of single-use products is not a new phenomenon due to the COVID-19 pandemic, but an agenda that has already been discussed for decades (Kalina & Tilley, 2020). Even with the crisis over, the trend of increasing waste generation is likely to continue.

The World Bank warned that "... looking forward, global waste is expected to grow to 3.40 billion tonnes by 2050, more than double population growth over the same period" (The World Bank, 2016, para. 2). Figure 15.1 illustrates the projected waste generation by region which has been increasing prior to the pandemic. This trend requires us to make efforts in the mid-and long-term rather than having short-term measures to curb the increase in waste generation. Without such efforts, the rapid growth of waste generation poses serious challenges for the planet. With the pandemic and growth of the stay-at-home economy, global plastic packaging usage is anticipated to further increase.

In line with the trend, the key question arises: How should companies and individual consumers approach to tackle the problem of increasing wastes generation? However, this question is not new. Many groups, including academia, international organisations, and non-governmental organisations have begun exploring this question. Such a concerted effort

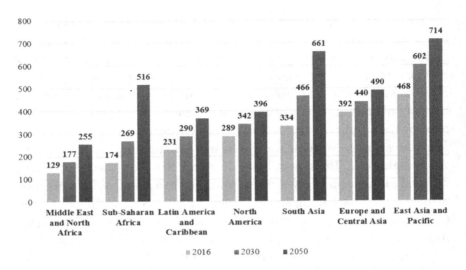

FIGURE 15.1 Projected waste generation, by region (millions of tonnes/year). (Source: Adapted by authors from The World Bank [2016])

has led to the idea of 'Sustainable Consumption and Production (SCP)'. The United Nations Environment Programme (UNEP) (2010) describes the background from which the concept of SCP was derived as follows:

> At the UN Conference on Environment and Development, held in Rio de Janeiro in 1992, sustainable consumption and production (SCP) was recognised as an overarching theme to link environmental and development challenges. The conference's final report, Agenda 21, states that the major cause of the continued deterioration of the global environment is the unsustainable patterns of consumption and production. ... At the World Summit on Sustainable Development (WSSD) in 2002 the Johannesburg Plan of Implementation was agreed, with a call for action to "encourage and promote the development of a 10-Year Framework of Programmes in support of regional and national initiatives to accelerate the shift towards sustainable consumption and production to promote social and economic development within the carrying capacity of ecosystems by delinking economic growth and environmental degradation."
>
> (pp. 5, 12)

Nonetheless, the consideration of pathways leading to a system for more sustainable consumption and production needed more attention

(Tukker et al., 2010). Circular economy approaches, in a broader context, can play a role for the transformation to systems of sustainable consumption and production (Schroeder et al., 2019). To that end, each member of society needs to have an urgent and collaborative commitment to circular economy approaches that can increase the efficiency of resource use in the production, distribution, and use of products within the carrying capacities of natural ecosystems during and after the COVID-19 pandemic. Without collaborative efforts to protect the environment, we are unlikely to achieve sustainable consumption and production (Adyel, 2020).

15.3 CHANGING UNSUSTAINABLE CONSUMPTION AND PRODUCTION PATTERNS

We are currently consuming more resources than ever before due to socioeconomic and demographic changes, exceeding the planet's capacity. Mapping out compatibility conditions between environmental sustainability and economic growth is one of the most significant global challenges. If we do not manage our current production and consumption patterns, the planet will need 183 billion tonnes of materials per year by 2050 (UNEP, 2020a). This is triple the amount of consumption today, which is almost impossible for the planet to handle. In order to respond to this situation, UNEP has been making global efforts over the past 20 years to transition to sustainable production and consumption patterns.

As part of its effort, Goal 12: Responsible Consumption and Production to ensure sustainable production and consumption patterns was established as 1 of the 17 UN SDGs in 2015. Sustainable consumption and production refer to "the use of services and related products, which respond to basic needs and bring a better quality of life while minimising the use of natural resources and toxic materials as well as the emissions of waste and pollutants over the life cycle of the service or product so as not to jeopardise the needs of future generations" (UNEP, 2020b, para. 4).

Why do sustainable consumption and production matter? The dilemma between the consumption-led growth and sustainability that we are currently facing may answer this question. Rapid industrialisation and globalisation that the world experienced have fueled our continued consumption of resources, especially in industrialised societies.

"Recent warnings confirm alarming trends of environmental degradation from human activity, leading to profound changes in essential life-sustaining functions of planet Earth" (Wiedmann et al., 2020, p. 1). In short, sustainability issues raised to this day warn us that our current consumption and production patterns in the guise of 'growth' can lead to severe damage to ecosystems, livelihoods, and our health.

Here, the idea of sustainable consumption and production comes into play. One of the sustainable consumption and production key goals is to 'decouple' economic growth and environmental degradation (UNEP, 2010). The goal is summed up by 'doing more and better with less impact' at all levels, which increases net welfare gains from economic activities while reducing resource use with sustainable lifestyles. Therefore, sustainable consumption and production can contribute to decoupling negative environmental impacts from the economic growth as a win-win solution for both the economy and the environment (O'Rourke & Lollo, 2015).

Above all, the successful transition to sustainable consumption and production requires a profound understanding of the entire life cycle of activities (European Union, 2010; Tseng et al., 2013) from the extraction of material, to design and production, packaging and finally to extended usage and end-of-life management. Figure 15.2 shows a schematic of a typical product life-cycle considering all the major phases from materials extraction to end-of-life management including reuse, recovery, recycling, and disposal of remaining components (Hao et al. 2020).

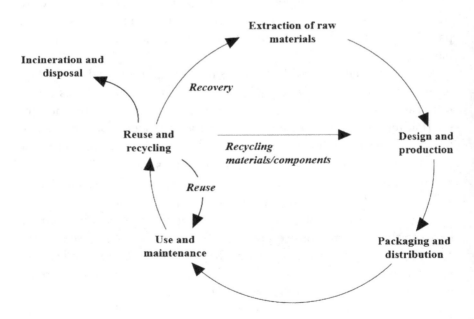

FIGURE 15.2   Major stages of a product life cycle. (Source: Adapted by authors from UNEP [2015])

In this chapter, we present the view that sustainable consumption and production are about systemic change and 'doing more and better with less impact' to decouple economic growth from environmental degradation. An approach based on product life cycles and using its supply chains can help increase resource efficiency along with both production and consumption phases, which can ultimately be a powerful lever for transition to a more sustainable society. Therefore, in order to achieve sustainable consumption and production, intensified efforts to understand and practice life-cycle thinking must be preceded by all actors in economic activity. Then, in the end, the following questions remain: *What roles can businesses and individual consumers contribute to sustainable consumption and production within a society?*

15.4 REVERSE LOGISTICS AND DIGITAL SERVITISATION: A SOLUTION FOR "NEW NORMAL"

When a product reaches its end-of-life (EoL) stage, it is sometimes returned to the producer for remanufacturing, recovery, or recycling using an existing logistics chain (Hao et al., 2020). Unlike forward logistics where raw materials flow through a variety of manufacturing and handling steps to reach the final consumer, this flow is called 'reverse logistics' where the existing supply chain is reversed from the consumer (Figure 15.3).

The core of circular economy models—Resource recycle and recovery, Remanufacturing, Product life extension, Sharing platforms, and Product as a Service—seek sustainability through restoration and optimisation of

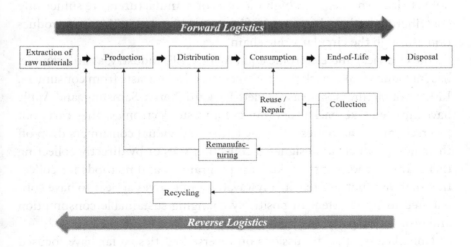

FIGURE 15.3   Forward and reverse logistics.

resources, away from one-time consumption. Reverse logistics provides an essential capability for the successful restoration of resources within the circular economy models. In other words, it contributes to increasing the efficiency of resources by enabling products or materials to be recycled, reused, and remanufactured in the value chain. Not only that, facilitating the implementation of reuse, repair, remanufacturing, and recycling also contributes to the transition of a linear value chain to a circular value chain (Ellen Macarthur Foundation, 2016). Therefore, reverse logistics help change unsustainable consumption and production patterns.

15.4.1 Reverse Logistics to Create New Values in the Loop

One point worth noting is that reverse logistics not only cover the collection, aggregation, and transport of products and materials, but also value-adding activities such as testing, sorting, separating, refurbishing, recycling, reprocessing, redistribution, and remarketing (Ellen Macarthur Foundation, 2016; Lacy et al., 2020). That is to say, a reverse logistics system goes beyond the meaning of mere 'product recovery' and includes the concept of creating opportunities for 'value discovery' through the value-adding activities.

Then, the question arises as to whether manufacturers are implementing reverse logistics beyond 'product recovery' today. Many advantages of reverse logistics, such as reducing the use of raw materials, increasing customer satisfaction, and protecting the environment, for example, have led manufacturers to pay more attention than ever in the past. However, it is not yet clear whether such high interests of manufacturers are sufficiently contributing to the achievement of 'sustainable consumption and production' through the circular value chain.

The electrical and electronic equipment (EEE) industry, as an example, has introduced take-back programmes to collect e-waste from consumers. EEE manufacturers, such as Hewlett-Packard, Xerox, Samsung, and Apple have initiated take-back programmes for waste electronics. They carry out reverse logistics activities either by letting individual consumers drop off their EoL products to designated collection sites or by directly collecting them. The purpose of the take-back programms and methods for collection may vary but as a result, anyway, their initiatives appear to have contributed to some extent in positively changing sustainable consumption and production patterns by closing the loop.

Nonetheless, still, discussions on reverse logistics so far have focused mainly on economic value for manufacturers (Islam & Huda, 2018;

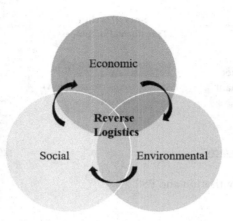

FIGURE 15.4   Three aspects of reverse logistics for a new circular economy.

Srivastava & Srivastava, 2006). An example is Xerox, an EEE manufacturer that saved US$10.2 million while eliminating 2.6 million pounds of waste from their take-back programme in 2010 alone (Kaye, 2010). Other values that reverse logistics can bring have received relatively little attention from academic and industry practitioners. Given that sustainability dimensions comprise economic, social, and environmental aspects (Islam & Huda, 2018), future reverse logistics practices need to be considered using economic, social and environmental dimensions (Figure 15.4) to create new values.

A new future circular economy will need a society-level recovery system that takes all parts into account (Esposito et al., 2018). To create new values within a reverse logistics system, we need to further expand our perspectives. We propose to explore new values through the value-adding activities within the reverse logistics system (see Section 15.3). By doing so, we will be able to create new values that encompass all social, environmental, and economic aspects. What is most encouraging is the fact that the potential of the value-adding activities is greatly increasing due to the development of information and communication technology (ICT). ICT will undoubtedly serve as a 'game changer' that facilitates the creation of new values within the reverse logistics system.

15.4.2 Digital Servitisation: The Aid to Close the Loop

ICT has enabled changes and improvements in reverse logistics, and also in supporting servitisation. The term servitisation was first introduced by Vandermerwe and Rada (1988) to a transformation process in which

FIGURE 15.5 Servitisation and PSS.

manufacturers create mutual values by shifting from selling products to selling Product-Service Systems (PSS), an integrated product and service offering (Figure 15.5). For example, servitised manufacturers provide customers with services ranging from product repair, monitoring, lease, renting, sharing, activity management to advice and consultancy related to products.

By introducing alternatives such as leasing, renting, and sharing, continuous production, and consumption of products may be altered and manufacturers can reduce the total use of resources (Kuah & Wang, 2020). Moreover, such scenarios also improve product usage efficiency by enabling individual consumers to use the product for a long time. Mont (2002), in his seminal paper, claimed the following benefits that PSS can bring to the environment.

> The PSS concept has the potential to bring about such changes in production and consumption patterns that might accelerate the shift towards more sustainable practices and societies … With PSS, producers become more responsible for their product–services in case material cycles are closed. Producers are encouraged to take back their products, upgrade and refurbish them and use them again. In the end, less waste is incinerated or landfilled.
>
> (pp. 239, 240)

Here, ICT can play a significant role in facilitating different types of service innovation. Digitally enabled offerings, such as remote monitoring or tracking devices to products, are allowing manufacturers to have new opportunities in the areas of maintenance, repair, and field operations, boosting servitisation through digitalisation (Coreynen et al., 2017). As the business environment becomes digitally oriented, manufacturers can leverage digital technologies such as IoT, Big Data, and Cloud to establish digital servitisation strategies that had better meet the needs of their customers.

The current crisis, triggered by the COVID-19 pandemic, limits the movement of human and material resources between or within countries. The unprecedented crisis calls for fundamental changes in manufacturers' traditional ways of producing and servicing. As the ongoing outbreak is unlikely to end in the near future (Lv et al., 2020), manufacturers should consider how to manage the current crisis and prepare for the "new normal" to come. We argue that a service strategy that makes extensive use of digital breakthroughs – a digital servitisation strategy – can pave the way for a new solution.

Digital servitisation has broadened the use of ICT, creating a new business strategy called 'smart-circular strategies': smart maintenance, smart reuse, smart remanufacturing, and smart recycling (Alcayaga et al., 2019; Zheng et al., 2019). Similarly, *green servitization strategies* that seek to achieve corporate goals for sustainability through the provision of green services (e.g. raw material recycling, maintenance, and repair services) have also been discussed in academia (De Silva et al., 2021; Marić & Opazo-Basáez, 2019; Opazo-Basáez et al., 2018). What they have in common is that they both used digital technologies to enable reverse logistics for firms.

As an example, a remanufacturing or recycling system supporting an EoL product can leverage digital capabilities. A system called 'smart recycling' improves the quality and results of product recycling by attaching sensors and product composition information to products themselves (Zheng et al., 2019). Such a system closes the loop on the product lifecycle while increasing the recycling operation efficiency. To that end, collaboration with external recyclers or specific IT units is an integral process required for efficient collection, sorting, and better disassembly. In the context of digital transformation, indeed, collaboration among various stakeholders is required to provide customers with customised solutions that require integrated supply chains. The collaboration between stakeholders that occurs in such an integrated supply chain becomes a driving force to create new values within the loop.

It is worth mentioning that Hewlett-Packard (HP) uses smart recycling as part of a digital servitisation strategy. HP Managed Print Services (MPS) and HP Instant Ink service are major 'product as a service' business models representing HP's circular economy initiative. These services help customers manage and optimise their printer fleets, ink cartridges, and digital workflows by servicing, maintaining, refurbishing, and redeploying units through cloud-based technology (HP, 2018). Cartridge recycling and returned cartridges are fed directly into a closed-loop recycling programme. As a result, approximately 75% of those printers are refurbished and remarketed. In doing so, HP has reduced the consumption of virgin materials in the production of new products (Dempsey, 2018) and achieved "a fully closed-loop recycling with HP-only streams through a collaborative effort with several key supply chain partners" (McIntyre & Ortiz, 2016, p. 322).

The HP's case is a good example of digital servitisation applied to the consumption and end-of-life phases for product remanufacturing and recycling. Reverse logistics activities are indispensable to achieve such a service process successfully. Here, the quality of logistics service can significantly increase with the adoption of digital technologies.

We expect extra values from digital adoption; more accurate tracking and more transparent real-time information of assets through IoT, Big Data, and Cloud technologies, for instance (Figure 15.6). Many scholars (Coreynen et al., 2017; Lerch & Gotsch, 2015; Martín-Peña et al., 2019), adopting a similar position, argued the development of digital technologies being an enabler of servitisation strategies.

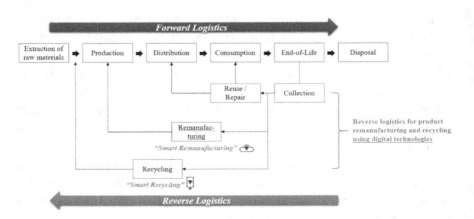

FIGURE 15.6   Digital servitisation as an aid to close the loop.

15.4.3 How to Implement the Reverse Logistics and Digital Servitisation in a Post-COVID World

In an increasingly digitised environment, we believe that companies can deliver more than ever values through the active introduction of digitised service strategies, such as smart remanufacturing and smart recycling. It provides an opportunity to reflect on values we have often overlooked so far in traditional sales and marketing, consumption, and even production activities. The biggest challenge we are facing is a transition to a society where sustainable consumption and production patterns can take root.

This is not a challenge caused solely by the COVID-19 pandemic, but the pandemic allows mankind to rethink, examine and reflect on their consumption patterns and wasteful habits. Many factors will help us achieve the goal, but we argued in this paper that reverse logistics and digital servitisation are key solutions for the 'new normal'. Putting the previous discussions together, Figure 15.7 below shows how reverse logistics and digital servitisation works together in the new circular economy in ensuring prosperity to humankind.

We discussed earlier a system of smart recycling where the quality and efficiency of recycling improves by attaching sensors and product composition information. HP's example of digital servitisation for product remanufacturing and recycling achieve success due to reverse logistics and the adoption of digital technologies. Extra value propositions had been achieved from digital adoption; more accurate tracking and more transparent real-time information.

The development and application of a digitised service platform to support this is paramount. Mobile applications, for example, can play

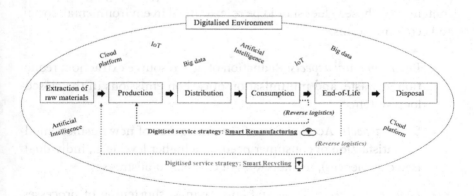

FIGURE 15.7   An illustration of reverse logistics and digital servitisation.

an important role in enabling companies to promote such strategies. Companies can use a service platform to invite various stakeholders (e.g. external suppliers, complementors, customers) to make a collaborative effort for value co-creation in the platform. Apple's platform business illustrates this. The range of digital servitisation adopted by Apple encompasses not only services directly related to Apple's physical products, but also services catered to different situations of Apple users. The opening of a platform such as iTunes or App Store, which provides content tailored to the needs of customers, is a service solution for Apple users. Apple has opened the App Store, allowing multiple service providers to deliver their software Apple users need. This shows an example in which several small platforms that provide solution services coexist with one large platform, iTunes or App Store. In doing so, external complementors and suppliers involved in Apple's platform business create competitive solutions by collaborating with Apple for their benefits.

In a similar vein, we contend that the implementation of smart remanufacturing/recycling strategies by other electronic manufacturers can contribute to the achievement of sustainable consumption and production to some extent by activating complementors running CE-related businesses within the industrial ecosystem, such as recyclers, remanufacturers, waste haulers. The business efforts of electronic manufacturers and complementors that they put in for their benefits, ironically, will be the driving force to change the pattern of sustainable consumption and production in a society.

Digital servitzation has a positive impact on value creation, value delivery, value capture, and value proposition. Although we agree to this, we contend that those values should be re-examined in environmental, social, and economic terms.

- *Environmental aspects:* Reduction of new resource extraction, reduction of waste, and minimisation of carbon emissions in the closed-loop system.

- *Social aspects:* Acceleration of the emergence of new business models, satisfaction of consumer needs at a higher level (e.g., individualised experiences), revitalisation of CE-related businesses.

- *Economic aspects:* Time and cost savings, shortening of processes, raising brand awareness as a green company.

Reverse logistics and digital servitisation can create new values in the three aspects above by playing a role as 'value re-discovery' beyond simple 'product recovery'. Such strategies will ultimately contribute to the establishment of sustainable consumption and production patterns in our society while meeting the three dimensions of sustainability. As an example, waste collection bins applied with the IoT technology will play an important role in creating a change in the pattern in which waste will be collected, sorted, reprocessed, and redistributed.

15.5 CONCLUDING REMARKS: TOWARDS A NEW CIRCULAR ECONOMY

In this chapter, we argue that reverse logistics and digital servitisation will be the driving force behind the transition to a new circular economy society where sustainable practices are prevalent. More importantly than that, we believe that when such actions are implemented taking into account environmental, social, and economic perspectives, the cornerstone for a new circular economy society will finally be laid. Figure 15.8 represents a comprehensive overview of our arguments and discussions in this chapter.

The aim of establishing sustainable consumption and production patterns in our society is not limited to reducing waste generation aggravated by the COVID-19 pandemic. It is important to know that we have already entered the 'Anthropocene' era. That is, if the current consumption and production patterns were to maintain their trajectories, our environment sustainability will not be guaranteed. This problem also relates to the survival of all the millions of different life forms that exist on our planet

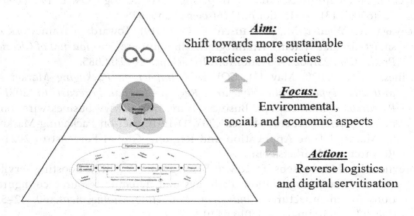

FIGURE 15.8   The shift towards sustainable consumption and production.

today, including mankind. The concept of circular economy, born to prevent such doom, is a novel economic system that deserves greater attention from all of us.

The future circular economy will need a society-level recovery system that would account for all its parts. We propose reverse logistics and digital servitisation can change the production patterns of companies and consumption patterns of individual consumers. Individual consumers should actively participate in corporate CE initiatives, thereby encouraging companies to take on more initiatives. Such collaborative efforts by companies and consumers will establish a sustainable consumption and production pattern, which will eventually prove that environmental degradation and economic development are compatible. Here, we reiterate that we should not pay attention only to the economic benefits from the virtuous cycle, but also to further environmental and social benefits.

Last but not least, we contend that policy support from governments can act as a catalyst to change the patterns of business and individual consumers. Governments should accurately diagnose the current situation based on their respective contexts and establish optimal policies for transition to a sustainable consumption and production society. Expecting voluntary changes in corporate and individual consumers' patterns without government policy support could be a cause of slowing the transition to a new circular economy society.

REFERENCES

Adyel, T. M. (2020). Accumulation of plastic waste during COVID-19. *Science, 369*(6509), 1314–1315. doi:10.1126/science.abd9925.

Alcayaga, A., Wiener, M., & Hansen, E. G. (2019). Towards a framework of smart-circular systems: An integrative literature review. *Journal of Cleaner Production, 221,* 622–634. doi:10.1016/j.jclepro.2019.02.085.

Business Wire. (2020, May 11). *COVID-19 Impact on Packaging Market by Material Type, Application and Region—Global Forecast to 2021— ResearchAndMarkets.com.* Business Wire. https://www.businesswire.com/news/home/20200511005497/en/COVID-19-Impact-on-Packaging-Market-by-Material-Type-Application-and-Region—Global-Forecast-to-2021—ResearchAndMarkets.com.

Coreynen, W., Matthyssens, P., & Van Bockhaven, W. (2017). Boosting servitization through digitization: Pathways and dynamic resource configurations for manufacturers. *Industrial Marketing Management, 60,* 42–53. doi:10.1016/j.indmarman.2016.04.012.

Dempsey, M. (2018). *Circular economy at HP: Bringing sustainability full circle*. Retrieved December 3, 2020, from ZeroWaste Scotland. https://www. zerowastescotland.org.uk/sites/default/files/ZWS%20CE%20Lecture%20 1%20Mark%20Dempsey%20HP%20Inc.pdf.

De Silva, M., Wang, P., & Kuah, A. T. H. (2021). Why wouldn't green appeal drive purchase intention? Moderation effects of consumption values in the UK and China. *Journal of Business Research, 122,* 713–724. https://doi. org/10.1016/j.jbusres.2020.01.016.

Donthu, N., & Gustafsson, A. (2020). Effects of COVID-19 on business and research. *Journal of Business Research, 117,* 284–289. doi:10.1016/j.jbusres. 2020.06.008.

Ellen Macarthur Foundation. (2016). *Capturing the value of the circular economy through reverse logistics: An introduction to the reverse logistics maturity model.* https://www.ellenmacarthurfoundation.org/assets/downloads/ ce100/Reverse-Logistics.pdf.

Esposito, M., Tse, T., & Soufani, K. (2018). Reverse logistics for postal services within a circular economy. *Thunderbird International Business Review, 60*(5), 741–745. doi:10.1002/tie.21904.

European Union. (2010). *Making sustainable consumption and production a reality: A guide for business and policymakers to life cycle thinking and assessment.* https://eplca.jrc.ec.europa.eu/uploads/LCT-Making-sustainable-consumption-and-production-a-reality-A-guide-for-business-and-policy-makers-to-Life-Cycle-Thinking-and-Assessment.pdf.

Hao, S., Kuah, A. T. H., Rudd, C. D., Wong, K. H., Lai, N. Y. G., et al. (2020). A circular economy approach to green energy: Wind turbine, waste, and material recovery. *The Science of the Total Environment, 702,* 135054–135054. https://doi.org/10.1016/j.scitotenv.2019.135054.

HP (Hewlett-Packard). (2018). *HP 2018 Sustainable Impact Report.* Available at: http://h20195.www2.hp.com/v2/GetDocument.aspx?docname=c06293935 (Accessed on 28 November 2020).

Islam, M. T., & Huda, N. (2018). Reverse logistics and closed-loop supply chain of waste electrical and electronic equipment (WEEE)/E-waste: A comprehensive literature review. *Resources, Conservation and Recycling, 137,* 48–75. doi:10.1016/j.resconrec.2018.05.026.

Kalina, M., & Tilley, E. (2020). "This is our next problem": Cleaning up from the COVID-19 response. *Waste Management, 108,* 202–205. doi:10.1016/ j.wasman.2020.05.006.

Kaye, L. (2010, November 15). *Xerox employees' green ideas save company $10.2 Million.* The Guardian. https://www.theguardian.com/sustainable-business/xerox-employees-green-ideas-save.

Kuah, A. T. H., & Wang, P. (2020). Circular economy and consumer acceptance: An exploratory study in east and Southeast Asia. *Journal of Cleaner Production, 247,* 119097. https://doi.org/10.1016/j.jclepro.2019.119097.

Lacy, P., Long, J., & Spindler, W. (2020). *The Circular Economy Handbook: Realizing the Circular Advantage.* Palgrave Macmillan, Springer Nature Limited.

Lerch, C., & Gotsch, M. (2015). Digitalized product-service systems in manufacturing firms: A case study analysis. *Research Technology Management*, *58*(5), 45–52. doi:10.5437/08956308x5805357.

Lv, H., Wu, N. C., & Mok, C. K. P. (2020). COVID-19 vaccines: Knowing the unknown. *European Journal of Immunology*, *50*(7), 939–943. doi:10.1002/eji.202048663.

Marić, J., & Opazo-Basáez, M. (2019). Green servitization for flexible and sustainable supply chain operations: A review of reverse logistics services in manufacturing. *Global Journal of Flexible Systems Management*, *20*(S1), 65–80. doi:10.1007/s40171-019-00225-6.

Markets and Markets. (2020). *COVID-19 Impact on Packaging Market by Material Type (Plastics/Polymers, Paper & Paperboard, Glass and Metal), Application (Healthcare, Food & Beverages, Household Hygiene, Beauty & Personal Care and Electrical & Electronics) and Region - Global Forecast to 2021*. Retrieved October 23, 2020, from Markets and Markets. https://www.marketsandmarkets.com/Market-Reports/covid-19-impact-on-packaging-market-27035953.html.

Martín-Peña, M., Sánchez-López, J., & Díaz-Garrido, E. (2019). Servitization and digitalization in manufacturing: The influence on firm performance. *Journal of Business & Industrial Marketing*, *35*(3), 564–574. doi:10.1108/JBIM-12-2018-0400.

McIntyre, K., & Ortiz, J. A. (2016). Multinational corporations and the circular economy: How Hewlett Packard scales innovation and technology in its global supply chain. In R. Clift & A. Druckman (Eds), *Taking Stock of Industrial Ecology* (pp. 317–330). Springer International Publishing.

Mont, O. K. (2002). Clarifying the concept of product–service system. *Journal of Cleaner Production*, *10*(3), 237–245. doi:10.1016/s0959-6526(01)00039-7.

Opazo-Basáez, M., Vendrell-Herrero, F., & Bustinza, O. (2018). Uncovering productivity gains of digital and green servitization: Implications from the automotive industry. *Sustainability*, *10*(5), 1524. doi:10.3390/su10051524.

O'Rourke, D., & Lollo, N. (2015). Transforming consumption: From decoupling, to behavior change, to system changes for sustainable consumption. *Annual Review of Environment and Resources*, *40*(1), 233–259. doi:10.1146/annurev-environ-102014-021224.

Schroeder, P., Anggraeni, K., & Weber, U. (2019). The relevance of circular economy practices to the sustainable development goals. *Journal of Industrial Ecology*, *23*(1), 77–95. doi:10.1111/jiec.12732.

Sharma, H. B., Vanapalli, K. R., Cheela, V. S., Ranjan, V. P., Jaglan, A. K., et al. (2020). Challenges, opportunities, and innovations for effective solid waste management during and post COVID-19 pandemic. *Resources, Conservation and Recycling*, *162*, 105052–105052. doi:10.1016/j.resconrec.2020.105052.

Srivastava, S. K., & Srivastava, R. K. (2006). Managing product returns for reverse logistics. *International Journal of Physical Distribution & Logistics Management*, *36*(7), 524–546. doi:10.1108/09600030610684962.

The World Bank. (2016). *Trends in solid waste management*. Retrieved November 01, 2020, from The World Bank. https://datatopics.worldbank.org/what-a-waste/trends_in_solid_waste_management.html.

Tseng, M., Chiu, A. S. F., Tan, R. R., & Siriban-Manalang, A. B. (2013). Sustainable consumption and production for Asia: Sustainability through green design and practice. *Journal of Cleaner Production*, *40*, 1–5. doi:10.1016/j.jclepro. 2012.07.015.

Tukker, A., Cohen, M. J., Hubacek, K., & Mont, O. (2010). Sustainable consumption and production. *Journal of Industrial Ecology*, *14*(1), 1–3.

United Nations Environment Programme. (2010). *ABC of SCP: Clarifying Concepts on Sustainable Consumption and Production*. https://sustainable development.un.org/index.php?page=view&type=400&nr=945&menu= 1515.

United Nations Environment Programme. (2015). *Sustainable consumption and production: A handbook for policymakers*. https://sustainabledevelopment. un.org/content/documents/1951Sustainable%20Consumption.pdf.

United Nations Environment Programme (UNEP). (2020a). Issue brief SDG 12: *Ensuring sustainable consumption and production patterns*. https://wedocs. unep.org/bitstream/handle/20.500.11822/25764/SDG12_Brief.pdf?sequence= 1&isAllowed=y.

United Nations Environment Programme (UNEP). (2020b). *Sustainable consumption and production policies*. Retrieved November 10, 2020, from United Nations Environment Programme. https://www.unenvironment.org/explore-topics/ resource-efficiency/what-we-do/sustainable-consumption-and-production-policies.

Vandermerwe, S., & Rada, J. (1988). Servitization of business: Adding value by adding services. *European Management Journal*, *6*(4), 314–324. doi:10. 1016/0263-2373(88)90033-3.

Wiedmann, T., Lenzen, M., Keyßer, L. T., & Steinberger, J. K. (2020). Scientists' warning on affluence. *Nature Communications*, *11*(1), 3107–3107. doi:10. 1038/s41467-020-16941-y.

World Economic Forum. (2020). *Is staying in the new going out? How the COVID-19 pandemic is fuelling the stay-at-home economy*. Retrieved October 21, 2020, from World Economic Forum. https://www.weforum.org/agenda/2020/ 05/covid19-coronavirus-digital-economy-consumption-ecommerce-stay-at-home-online-education-streaming/.

Zheng, P., Wang, Z., Chen, C., & Pheng Khoo, L. (2019). A survey of smart product-service systems: Key aspects, challenges and future perspectives. *Advanced Engineering Informatics*, *42*, 100973. doi:10.1016/j.aei.2019.100973.

CHAPTER 16

Agriculture and the Future of Food through Digital and Other Novel Technologies

Stewart Lockie[1,2]

[1]*James Cook University Singapore, Singapore*
[2]*The Cairns Institute, Cairns, Australia*

CONTENTS

DOI: 10.1201/9781003148715-16

16.1 INTRODUCTION

The case to transform agriculture is compelling. Climate change, natural resource degradation, population growth and urbanisation challenge both, the wellbeing of rural communities and the adequacy, stability, safety and quality of global food supplies (Lockie 2015). While consumer demand for high quality, socially and environmentally responsible products creates new market opportunities, agriculture-dependent communities worldwide remain among the most poverty-stricken and food insecure. Rising to these challenges demands transformational change in the productivity and climate resilience of farming systems, in the efficiency and equity of agricultural supply chains, and in the human and social capital available to rural communities (Fleming et al. 2018; Lockie et al. 2020).

The potential of so-called Fourth Industrial Revolution technologies to facilitate agricultural transformation also appears compelling (Klerkx et al. 2019). Novel digital, bio- and nanotechnologies afford a dizzying array of possibilities to resolve problems within existing production and distribution systems and to develop entirely new products, processes and markets. Integrating advanced technologies affords even more possibilities (Lockie et al. 2020). From the development of crops and livestock with enhanced nutrient profiles, water use efficiency, disease resistance, and so on, to the deployment of advanced sensing, data analytics and artificial intelligence to optimise labour, natural resource and capital inputs and the use of microbes to produce high value medicinal and food products, the future of agriculture appears to be one in which profitability and the goals of poverty alleviation, food security, environmental protection and climate adaptation are increasingly achievable. Indeed, if Sustainable Development Goals agreed by member states of the United Nations in 2015 – goals that include ending poverty and hunger while de-coupling economic growth from rising resource consumption – are to be realised, it appears that digital and other novel agricultural technologies are an essential part of the solution.

It is worth bearing in mind, however, that every technological revolution through history has been, at the same time, an institutional, political, social and cultural revolution (Lockie and Wong 2018). Advanced technologies offer tantalising possibilities to increase productivity, commercialise new products, differentiate produce, verify provenance and quality, manage climate and biosecurity risks, solve environmental problems, improve animal welfare and lift the profitability of agriculture. The specific ways in which these possibilities materialise will depend though in no small way on how governments, researchers, entrepreneurs, farmers and other businesses navigate a range of distinctly social and political issues. Alternative possibilities, after all, include the concentration of market power, displacement of smallholders and agricultural workers, and declines in agricultural biodiversity that have characterised previous waves of technological innovation. Regulatory frameworks, trade- and geopolitics, research and development outcomes, corporate strategy, investment markets, civil society movements and consumer preferences will all have roles to play in shaping the outcomes of this most recent technological revolution.

The disruptive potential of digital and other novel technologies is obvious. The question is how this potential can be harnessed to pursue outcomes that are genuinely transformational from the perspective of sustainable and inclusive development. This chapter will explore the promise of advanced technologies in a little more detail before turning to risks and opportunities for ensuring prosperous and equitable food futures.

16.2 THE PROMISE OF DIGITAL, BIO- AND NANOTECHNOLOGIES FOR AGRICULTURE

The Fourth Industrial Revolution, according to the World Economic Forum,[1] is characterised by technological innovation of a speed and complexity for which there is no historical precedent, exponential increases in computational capacity and data capture, the convergence of digital and biotechnologies and consequently, a merging of the physical, biological and digital worlds. A plethora of technologies is implicated in this revolution. Nonetheless, Lockie et al. (2020) identified those both most likely both to impact (Australian) agriculture within the next ten years and most relevant to major challenges facing the sector as:

- sensors and the Internet of Things
- automation management technologies (robotics, machine learning, large-scale optimisation and data fusion)
- synthetic biotechnology
- nanotechnology
- transactional technology (distributed ledger and personalisation technologies)

Sensor technology offers myriad opportunities for detecting events or changes in the surrounding environment relevant to biosecurity, plant and animal health, soil and water management, food safety, and so on. Networking sensors through the Internet of Things (IoT) facilitates the use of real-time information for monitoring and decision-making.

Intelligent use of sensor and IoT data requires the synthesis and extraction of functionally useful information. Machine learning allows automated systems to process and adapt independently to large quantities of data, allowing them to undertake tasks more commonly associated with the exercise of human intelligence. Data fusion and large-scale optimisation, on the other hand, draw together multiple sources of data and apply mathematical techniques designed to address problems too large and/or complex for standard optimisation methods.

Sensors, IoT, artificial intelligence, data fusion and large-scale optimisation will support a range of capabilities more specific to agriculture including crop monitoring, process automation, robotics and decision-making. Coupled with improved weather and market forecasting, these capabilities will support enhanced risk management, asset optimisation and labour utilisation. In principle, sensors, robotics and automation will improve occupational health and safety and allow agricultural workers to devote more time to complex tasks.

Synthetic biology refers to the design and construction of artificial biological pathways, organisms, networks or devices, or the redesign of biological systems (Gray et al. 2018). Of particular interest are: (1) *RNA-interference* techniques used to silence genes which could be used to reduce undesirable plant or animal characteristics such as the production of toxins; (2) *gene-editing* techniques such as CRISPR-Cas9 which, unlike genetic modification techniques that transfer DNA between species, manipulate only one or a few nucleotides of the DNA sequence to breed organisms with more desirable traits; and (3) faster and less costly *gene sequencing* techniques that facilitate increased understanding of biological systems, more efficient plant and animal breeding, and the detection of pathogens and pathogen characteristics such as antibiotic resistance. All three of these techniques can be used to develop organisms and production systems with increased drought tolerance, pest or disease resistance, productivity or product quality.

Nanotechnologies utilise materials that, by virtue of their extremely small size, exhibit unique physical and chemical properties such as large surface areas, high solubility, catalytic reactivity, or specific sizes and shapes. Nanomaterials are defined as those in the range of 1 to 100 nano-metres for at least one of three dimensions, or roughly 800 to 100,000 times smaller than the diameter of a human hair. Nanotechnology applications in agriculture are currently rare. However, the scope of potential applications is vast – ranging from the use of plants to grow nanoparticles for drug delivery to the development of biosensors for environmental monitoring, medical and veterinary diagnosis and drug discovery, and the nano-formulation of more targeted fertilisers and pesticides (Gray et al. 2018).

Distributed ledger technologies (DLT) record and update data structures, or ledgers, simultaneously in multiple places. Blockchain, the most advanced and widely used DLT, utilises public key cryptography, peer-to-peer networking, databases, game theory, and consensus algorithms to record and track information in a shared, distributed and decentralised manner, facilitating transactions and value transfer. Initially developed as a cryptocurrency platform, blockchain technology has been applied to the problem of establishing supply chain integrity most notably, perhaps, with respect to the global diamond trade. Given the importance to agriculture of verifiable quality standards, product traceability and supply chain transparency (Botterill and Daugbjerg 2011; Lockie 2020), blockchain and other DLTs are seen as promising avenues to reduce the cost of managing transactional data and to use these data more effectively to build both trust and value (Kamilaris et al. 2019).

Each of the technologies discussed above can be integrated within multiple higher-level applications and capabilities. While single technologies may be used to improve the efficiency or effectiveness of discrete farming and supply chain practices, combinations of technology (old and new) are more likely needed to support the development of wholly new systems and products (Kirkegaard 2019). Personalisation technologies offer a case in point. Mass personalisation technologies that utilise customer information, data analytics, machine learning and artificial intelligence to customise marketing have profoundly reshaped, already, the retail and distribution landscapes. In contrast, however, with the concentration of market power associated with mass personalisation, advanced technologies can also be used to support the diversification of production and distribution systems. More direct relationships between producers and consumers, the differentiation of agricultural products according to both tangible and intangible product attributes (texture and taste, for example, versus safety, cultural heritage and social and environmental responsibility), possibilities to develop new products with specific functional attributes (such as nutraceuticals) and cost-effective systems to verify product claims have potential to deliver far more value directly to farmers and rural workers.

Alternative attempts to identify technologies with the potential to influence agriculture over the next decade or so flag a large number of additional candidates – augmented and extended reality, quantum computing, 5G, high-altitude wireless delivery, edge computing, acellular agriculture, human augmentation, 3D printing, drone delivery, perovskite solar cells, solar retransmission, artificial photosynthesis, programmable materials, and more (Hamilton et al. 2019). The further we move from technologies that are mature and/or technologies that can be considered foundational (in the sense they underwrite other technological capacities) the less certain we can be about what will eventually find its way into widespread commercial application. More importantly, the further we move from foundational technologies the less certain we can be about how technologies will be combined to create new agricultural products, production systems, business models and supply chain arrangements. Anticipation of these possibilities, however, is critical if a transformative agenda for agriculture is to be established (Fleming et al. 2018).

16.3 RISK AND THE ADOPTION OF NOVEL TECHNOLOGIES

The transformative potential of digital, bio- and nanotechnologies has stimulated a burst of activity among governments, agri-industry associations and research institutions seeking to anticipate technological developments and accelerate their application in agriculture (Hamilton et al. 2018, 2019; Klerkx et al. 2019; Leonard et al. 2017; Lockie et al. 2020). This is driven as much by the opportunities afforded by novel technologies as it is by fear of falling behind competitors, failing to deal with the environmental, economic and social challenges facing agriculture and failing further, importantly, to mitigate emerging risks.

Fear of falling behind positions farmers and/or consumer resistance to the adoption of particular technologies among the most immediate risks for attention. Dealing with consumer resistance first, while there are limits to what can be extrapolated from reactions to other rounds of technological innovation in agriculture there are important lessons to be drawn, nonetheless, from resistance to the commercialisation of genetically modified organisms (GMOs) over the last thirty or so years.

16.3.1 Consumer Resistance

Research shows that consumers, broadly, are most concerned about the genetic modification of animals, of products they ingest, and products that offer them no tangible benefits (Lockie et al. 2005). They are less concerned about the modification of fibre, ornamental and/or medicinal products. Further, women are generally more concerned about genetic modification than are men, as are those with lower levels of education (Bray and Ankeny 2017; Lockie et al. 2005), although the provision of more information about genetic modification can amplify opposition (Lawrence et al. 2001). Concerns about genetic modification reflect preferences for foods perceived as traditional or natural and the association of these with other values including caring for others and caring for nature (Lockie et al. 2005). Concerns about genetic modification also, importantly, reflect a lack of trust in agribusiness firms marketing GMOs and in regulatory institutions responsible for protecting consumer interests (Bearth and Siegrist 2016). Consumers have questioned whether GMOs are as safe and environmentally benign as their proponents claim and they have questioned, perhaps more critically, the motivations of those proponents. A contradiction has been evident, in the minds of many, between the promotion of genetic engineering as a solution to everything from poverty and food insecurity to environmental degradation and animal cruelty and the priority accorded to clearly production-focused traits such as herbicide resistance in the development of GMOs for commercial release (Hindmarsh and Lawrence 2001).

We should not expect all novel technologies to elicit active consumer resistance and indeed novelty is itself a valued attribute among many consumers. The GMO experience suggests not that agriculture should remain trapped in a technological time warp but that: (1) novelty and innovation should be directed to the interests and concerns of consumers; and (2) claims to deliver broader societal and environmental benefits should be realistic and demonstrable. Where this is achieved there is potential to add significant economic value to agricultural products as evidenced by substantial retail price margins on organic, fair trade and other foods that are seen to embody defensible claims of social and environmental responsibility (Lockie et al. 2006). And where this is not achieved there is a significant risk of buyer backlash at multiple scales as evidenced by the widespread imposition of retailer-defined agricultural production standards in addition to consumer boycotts (Lockie 2020).

16.3.2 Farmer Adoption

Consistent agricultural productivity growth since the 1950s and evidence that agriculture in advanced economies, such as Australia, is closing on the productivity frontier of existing technologies suggests that on-farm innovation is a well-established norm (Daly et al. 2015). At the same time, high capital costs, complexity, incompatibility with existing practices, unproven benefits, lack of institutional or technical support, and inconsistency with local knowledge and occupational identities are all known to slow adoption (Carruthers and Vanclay 2012; Guerin 1999; Mankad 2016; Pannell et al. 2006; Wheeler 2018). Applying these lessons to digital agriculture, Robertson (2018) usefully articulates a farmer-centric approach that stresses the development of systems that minimise steps between data collection and the extrapolation of useful information, help farmers test and improve their own knowledge, and allow farmers to combine data from digital systems with other sources of information.

As of now, however, there is evidence of scepticism among farmers toward digital agriculture in Australia and the United States that reflects uncertainty over who owns, accesses and controls data, how privacy is managed and who actually benefits (Carolan 2020; Jakku et al. 2019; Keogh and Henry 2016; Shepherd et al. 2018). Conversely, one study found that many more women than men utilise digital innovations in the Australian beef industry and find use of these innovations empowering (Hay and Pearce 2014). Similar research in other sectors, other countries, and even at other times, will elicit different results – attitudes toward emerging technologies tending to be fluid and uneven until those technologies mature and their real costs and benefits become clear (Skinner 2018). Farmers anticipate digital, automated futures but are not entirely sure what this will mean in context of the unique agro-ecologies, business structures and institutional environments in which they operate (Legun and Burch 2021). As the costs and benefits of specific digital agriculture applications become clearer, variance in attitudes will likely decrease and the availability of technical, legal and training infrastructure to support adopters relatively more important (Skinner 2018).

It is worth bearing in mind here that farmers may adopt what become widely used technologies despite reservations about their economic or environmental sustainability (Lockie et al. 1995). Departure from accepted practice and emerging technological trajectories carries risk as such departure is accompanied, often, with loss of access to peer and institutional support networks, readily available technical services and input suppliers, and so on. Despite the almost infinite number of ways in which advanced technologies could be operationalised many corporate actors are prioritising opportunities that allow them to protect and extend their own intellectual property rights while capturing, controlling and profiting from farm-level data (Carbonell 2016; Carolan 2020; Prause et al. 2020). The reservations many farmers express towards this kind of trajectory may dissipate over time. Equally, they may foster support for different approaches to digital agriculture (for example, open-source platforms[2] and data cooperatives), rejection of some digital innovations altogether, or resignation over a perceived lack of choice (see also Fleming et al. 2018).

There are, of course, farmers who embrace experimentation with unproven technologies just as there are many examples of complex and capital-intensive innovations that have, eventually, become commonplace. The point here is that early adoption always carries additional risk while particularly novel innovations may also confront potential adopters with a deficit of locally available technical and warranty support, uncertainty over costs and benefits, lack of familiarity with the underlying technologies, difficulty finding appropriately skilled employees, and regulatory frameworks that are not fit-for-purpose. Each of these challenges are magnified for smallholder and resource-poor farmers. Reductions in cost and the increasing capability of sensors, computational hardware, etc., can be expected to accelerate the adoption of advanced technology but it is also important to consider here issues that extend beyond the farm gate including provisions for the ownership and security of data, licensing costs for access to intellectual property, and the availability of robust telecommunications and data processing infrastructure.

16.3.3 Incremental versus Transformative Adoption

For all their transformative potential, the vast majority of field applications of advanced technologies in agriculture, to date, offer incremental improvements to existing farming systems (Kirkegaard 2019). For example, sensors and IoT technologies are being used to optimise water use efficiency in irrigation systems (see Table 16.1). Precision agriculture practices are integrating sensors, automation and global positioning systems to improve fertiliser and pesticide use efficiency. Gene sequencing is being used to speed up genetic improvement through otherwise conventional plant and animal breeding.

TABLE 16.1   Adaptive Responses to Environmental Change

Adaptive Response	Systemic Impact	Agricultural Examples
Incremental	Optimisation of existing systems	Monitoring (e.g. sensors, IoT)
		Forecasting (e.g. enterprise analytics, AI)
		Efficiency (e.g. automation, telecommunications)
	Substitution of system components	Genetic improvement (e.g. gene sequencing)
		Enhanced inputs (e.g. nanomaterials)
	Conversion to different systems	Enterprise change (e.g. land conversion)
Transformative	Incubation of novel systems	New food and fibre industries (e.g. acellular meat)

Source: Adapted from Lockie et al. (2020).

There is nothing inherently wrong with the optimisation of production systems (or distribution and processing systems for that matter) given even the most incremental rates of change result in profound and systemic transformation if sustained long enough (Kirkegaard 2019). However, the risks of incrementalism are twofold. First, change in agriculture may not be fast enough to keep pace with the escalating impacts of global environmental change. Increasing climate volatility and extreme weather event frequency will challenge the resilience not only of farming systems but of biosecurity and public health systems, supply chain relationships, distribution infrastructure, credit and insurance markets and urban food security. Second, investment in capital intensive technologies and infrastructure may prove maladaptive, over the medium- to long-term, if farmers, governments and other investors do not take adequate account of escalating climate impacts and potential for market disruption when making investment decisions.

16.4 AN ENVIRONMENT FOR TRANSFORMATIONAL CHANGE

Accelerating transformational change in agriculture is not a simple matter of accelerating the adoption of advanced digital, bio- and nanotechnologies. Barriers to the adoption of these technologies have certainly been identified and strategies to remove some these barriers will be discussed in this section. However, little of the technology discussed throughout this chapter is field ready and most is best understood, in any case, as a platform capability for the development of more applied technologies and farm management practices. Established areas of agricultural research, such as farming systems research, will retain their importance as they are used to help define the problems to which novel technological capacities might be applied and as they are used, subsequently, to establish the business case for novel solutions (Shepherd et al. 2018).

Accelerating transformational change in agriculture is also not a matter of focusing solely on production. Technology applications that support new business practices, supply chain relationships, value-added products and connections with consumers may be just as, if not more, important given the need to sustain businesses and communities through increasingly frequent and severe adverse climate events. Data analytics and other digital platforms offer interesting new opportunities to understand consumer demand and connect producers with consumers but their application must be extended beyond targeted marketing if the transformative potential of these platforms is to be fully realised. Consumer activists (or 'prosumer' movements) have had considerable influence already in relation to standards development, retail strategy, the growth of alternative food markets and, critically, the escalation of conflict over food safety, genetic modification, animal cruelty, labour rights and environmental performance (Lockie 2020). Finding ways to engage consumers, along with farmers and other stakeholders, in democratic dialogue over the future of agriculture will go a long way to avoiding boycotts over unwanted or untrusted technologies.

Dialogue takes time and its outcomes can be surprising. In the meantime, there are at least three kinds of infrastructure – physical, regulatory and institutional – that are themselves critical to the development of an environment supportive of transformational approaches to sustainable agricultural development.

16.4.1 Physical Infrastructure

Connectivity is fundamental. As technology changes the requirements of information and telecommunications systems will also change as will the costs of delivering these requirements. As of now, rural businesses face the same challenges with digital connection as they do with roads and other transport connections – systems that do not match those in urban centres and which are not capable of supporting emerging business needs (Mark et al. 2016; Shepherd 2018). The problem is particularly acute in the Global South but in no way is it restricted to the Global South. At issue, further, is not only bandwidth and network speed but access to appropriate data management infrastructure. The basic premise of edge computing – that storing and processing data where they are gathered and consumed improves response times while reducing bandwidth requirements – is thus as relevant to the vast rural landscapes in which extensive agriculture is practiced as it is to more obviously data-intensive innovations such as vertical farming. Regional data centres and other decentralised infrastructure have the potential, moreover, to support diversification of rural employment markets and the availability of skilled employees, facilitating further technology adoption.

16.4.2 Regulatory Infrastructure

Regulatory frameworks that protect the interests of innovators, end-users and the broader public are needed to reduce investment risk and encourage uptake. Implementation of effective regulatory frameworks is challenged by the novelty of advanced technology applications and by the complex, often global, networks of intellectual property holders, manufacturers and service providers involved in discrete technology applications. The peculiarities of each jurisdictional context aside, areas that may require regulatory development or reform to support innovation and adoption include (see Keogh and Henry 2017; Lockie et al. 2020):

- Data ownership, sharing and privacy
- Data standards including standards for open data, data exchange and related software
- Liability for errors or the misinterpretation of data by automated systems
- Commercialisation and intellectual property
- Safety standards, safeguards and labelling
- Regulatory approval for modified organisms, nanomaterials, etc.

It is important to note too that novel technologies may support new approaches to regulation elsewhere in the agricultural sector. For example, biosensors could be developed to monitor compliance with food safety or environmental regulation. Whether our focus is on regulation *of*, or *with*, new technologies, legitimacy and effectiveness are likely to depend on transparent consideration of all stakeholder interests.

16.4.3 Institutional Infrastructure

Transformative solutions to complex problems are likely to require the application of multiple technological platforms and capacities, the participation of diverse stakeholder groups and critically, dynamic responses to challenges and insights that emerge throughout the innovation cycle. As with regulatory infrastructure, unique jurisdictional circumstances will dictate unique institutional arrangements. Nonetheless, institutional infrastructure and governance arrangements that facilitate cross-sectoral and cross-disciplinary collaboration in research, development and innovation is far more likely to support transformational change than is institutional infrastructure which reinforces existing sectoral (or industry, departmental and social) boundaries (Shepherd et al. 2018). Lockie et al. (2020) argue that well-designed institutional arrangements have the potential to support virtuous cycles of locally relevant research and innovation, education, industry application, business diversification, economic development and social inclusion in agricultural regions.

16.5 CONCLUSION

Automation, the Internet of Things, data analytics and artificial intelligence are already finding their way onto farms, as are the products of synthetic biology and other emerging technologies. Agriculture has always been at the forefront of technological innovation and we can be absolutely certain that farms and agribusiness will remain key sites for the application of novel digital, bio- and nanotechnologies. What we can be rather less certain about is what this will mean in practice. How will technologies be deployed? What will be produced? What will be the economic, social and environmental outcomes? Who will benefit? The potential for sustainable and inclusive development is real but so too is the potential for concentration of intellectual property ownership, rationalisation of farm ownership, de-skilling of farmers and farmworkers, contraction of rural labour markets, degradation of water, soil and biodiversity resources, and failure to manage climate and biosecurity risks. The point here is not to paint the future as dystopian but to stress that no particular future is inevitable. If the potential of advanced technologies to support positive outcomes is to be realised it is incumbent on governments and other stakeholders to ensure that the institutional, regulatory and physical infrastructure to support positive and transformational change is in place. The problem, as articulated above, is not one of accelerating technology adoption but of enabling meaningful participation and collaboration among multiple stakeholder groups including, but not limited to, farmers, researchers, agribusinesses, technology firms and civil society organisations.

NOTES

1. www.weforum.org/focus/fourth-industrial-revolution. Accessed 13 March 2021.
2. Open access data arrangements are currently becoming more common in the United States, despite attempts by machinery manufactures and service providers to establish closed proprietary systems, as more competitive markets for software, data storage and data management emerge (Keogh and Henry 2016).

REFERENCES

Bearth, A. and Siegrist, M. (2016) Are risk or benefit perceptions more important for public acceptance of innovative food technologies: a meta-analysis. *Trends in Food Science & Technology* 49: 14–23.

Botterill, L. and Daugbjerg, C. (2011) Engaging with private sector standards: a case study of GLOBALG.A.P. *Australian Journal of International Affairs* 65(4): 488–504.

Bray, H. and Ankeny, R. (2017) Not just about 'the science': science education and attitudes to genetically modified foods among women in Australia. *New Genetics and Society* 36(1): 1–21.

Carbonell, I. (2016) The ethics of big data in agriculture. *Internet Policy Review*, 5(1): 1–13.

Carolan, M. (2020) Digitization as politics: smart farming through the lens of weak and strong data. *Journal of Rural Studies*. doi.org/10.1016/j.jrurstud. 2020.10.040.

Carruthers, G. and Vanclay, F. (2012) The intrinsic features of environmental management systems that facilitate adoption and encourage innovation in primary industries. *Journal of Environmental Management* 110: 125–134.

Daly, J., Anderson, K., Ankeny, R., Harch, B., Hastings, A. et al. (2015) *Australia's Agricultural Future*. Australian Council of Learned Academies, Melbourne.

Fleming, A., Lim-Camacho, L., Taylor, B., Thorburn, P. and Jakku, E. (2018). Is big data for big farming or for everyone? Perceptions in the Australian grains industry. *Agronomy for Sustainable Development* 38(3): 24.

Gray, P., Meek, S., Griffiths, P., Trapani, J., Small, I. et al. (2018) *Synthetic Biology in Australia: An Outlook to 2030*. Australian Council of Learned Academies, Melbourne.

Guerin, T. (1999) An Australian perspective on the constraints to the transfer and adoption of innovations in land management. *Environmental Conservation* 26(4): 289–304.

Hamilton, G., Swann, L., Kutty, S., Hearn, G., Nayak, R., et al. (2018) *Detecting Opportunities and Challenges for Australian Rural Industries: Final Report.* AgriFutures Australia, Wagga Wagga, NSW.

Hamilton, G., Swann, L., Pandey, V., Moyle, C., Opie, J. et al. (2019) *Horizon Scanning: Opportunities for New Technologies and Industries.* AgriFutures Australia, Wagga Wagga, NSW.

Hay, R. and Pearce, P. (2014) Technology adoption by rural women in Queensland, Australia: Women driving technology from the homestead for the paddock. *Journal of Rural Studies* 36: 318–327.

Hindmarsh, R. and Lawrence, G. (2001) Bio-utopia: future natural. In: Hindmarsh, R. and Lawrence, G. (Eds.), *Altered Genes II: the Future?* Scribe, Melbourne.

Jakku, E., Taylor, B., Fleming, A., Mason, C., Fielke, S., et al. (2019) 'If they don't tell us what they do with it, why would we trust them?' Trust, transparency and benefit sharing in Smart Farming. *NJAS – Wageningen Journal of Life Sciences* 90–91: 100285.

Kamilaris, A., Fonts, A. and Prenafeta-Boldú, F. (2019) The risk of blockchain in agriculture and food supply chains. *Trends in Food Science and Technology* 91: 640–652.

Keogh, M. and Henry, M. (2016). *The Implications of Digital Agriculture and Big Data for Australian Agriculture: April 2016.* Australian Farm Institute, Sydney, Australia.

Kirkegaard, J. (2019) Incremental transformation: success from farming system synergy. *Outlook on Agriculture* 48(2): 105–112.

Klerkx, L., Jakku, E. and Labarthe, P. (2019) A review of social science on digital agriculture, smart farming and agriculture 4.0) new contributions and future research agenda. *NJAS – Wageningen Journal of Life Sciences* 90–91: 100315.

Lawrence, G., Norton, J. and Vanclay, F. (2001) Gene technology, agri-food industries and consumers. In: Hindmarsh, R. and Lawrence, G. (Eds.), *Altered Genes II: The Future?* Scribe, Melbourne.

Legun, C. and Burch, K. (2021) Robot ready: how apple producers are assembling in anticipation of new AI robotics. *Journal of Rural Studies* 82: 380–390.

Leonard, E., Rainbow, R., Tridall, J., Baker, I., Barry, S., et al. (2017). *Accelerating Precision Agriculture to Decision Agriculture: Enabling Digital Agriculture in Australia*, Cotton Research and Development Corporation, Australia.

Lockie, S. (2015) *Australia's Agricultural Future: The Social and Political Context.* Report to SAF07 – Australia's Agricultural Future Project. Australian Council of Learned Academies, Melbourne.

Lockie, S. (2020) *Failure or Reform? Market-Based Policy Instruments for Sustainable Agriculture and Resource Management*, Routledge, London.

Lockie, S., Fairley-Grenot, K., Ankeny, R., Botterill, L., Howlett, B., et al. (2020) *The Future of Agricultural Technologies*, Australian Council of Learned Academies, Melbourne.

Lockie, S., Lawrence, G., Lyons, K. and Grice, J. (2005) Natural foods and biotechnologies: a path analysis of factors underlying support or opposition to biotechnology among Australian food consumers. *Food Policy* 30: 399–418.

Lockie, S., Lyons, K., Lawrence, G. and Halpin, D. (2006) *Going Organic: Mobilising Networks for Environmentally Responsible Food Production*, CABI Publishing, Wallingford, UK.

Lockie, S., Mead, A., Vanclay, F. and Butler, B. (1995) Factors encouraging the adoption of more sustainable cropping systems in South-East Australia: profit, sustainability, risk and stability. *Journal of Sustainable Agriculture* 6(1): 61–79.

Lockie, S. and Wong, C. (2018) Conflicting temporalities of social and environmental change? in Bostrom, M. and Davidson, D. (eds) *Environment and Society: Concepts and Challenges*, Palgrave Macmillan, Gewerbestrasse, CHE.

Mankad, A. (2016) Psychological influences on biosecurity control and farmer decision-making. A review. *Official Journal of the Institut National de La Recherche Agronomique (INRA)* 36(2): 1–14.

Mark, T., Griffin, T. and Whitacre, B. (2016) The role of wireless broadband connectivity on "big data" and the agricultural industry in the United States and Australia. *International Food and Agribusiness Management Review* 19(A): 43–56.

Pannell, D., Marshall, G., Barr, N., Curtis, A., Vanclay, F., et al. (2006) Understanding and promoting adoption of conservation practices by rural landholders. *Australian Journal of Experimental Agriculture* 46(11): 1407–1424.

Prause, L., Hackfort, S. and Lindgren, M. (2020) Digitalization and the third food regime. *Agriculture and Human Values*. doi.org/10.1007/s10460-020-10161-2.

Shepherd, M., Turner, J., Small, B. and Wheeler, D. (2018) Priorities for science to overcome hurdles thwarting the full promise of the 'digital agriculture' revolution. *Journal of the Science of food and Agriculture* 100: 5083–5092.

Skinner, A. (2018) Big data maturity in Australian agricultural industries. *Farm Policy Journal* 15(1).

Wheeler, S. (2008) What influences agricultural professionals' views towards organic agriculture? *Ecological Economics* 65(1): 145–154.

Index

Note: Locators in *italics* represent figures and **bold** indicate tables in the text.

Printed in the United States
by Baker & Taylor Publisher Services

Printed in the United States
by Baker & Taylor Publisher Services